Controlled Flight into Terrain (CFIT/CFTT)

The McGraw-Hill *CONTROLLING PILOT ERROR* Series

Weather
Terry T. Lankford

Communications
Paul E. Illman

Automation
Vladimir Risukhin

Controlled Flight into Terrain (CFIT/CFTT)
Daryl R. Smith

Training and Instruction
David A. Frazier

Checklists and Compliance
Thomas P. Turner

Maintenance and Mechanics
Larry Reithmaier

Situational Awareness
Paul A. Craig

Fatigue
James C. Miller

Culture, Environment, and CRM
Tony Kern

CONTROLLING PILOT ERROR

Controlled Flight into Terrain (CFIT/CFTT)

Daryl R. Smith, Ph.D.

McGraw-Hill

New York   Chicago   San Francisco   Lisbon   London   Madrid
Mexico City   Milan   New Delhi   San Juan   Seoul
Singapore   Sydney   Toronto

18.00

Cataloging-in-Publication Data is on file with the Library of Congress

McGraw-Hill

A Division of The McGraw-Hill Companies

Copyright © 2001 by The McGraw-Hill Companies, Inc. All rights reserved.
Printed in the United States of America. Except as permitted under the United
States Copyright Act of 1976, no part of this publication may be reproduced or
distributed in any form or by any means, or stored in a data base or retrieval
system, without the prior written permission of the publisher.

1 2 3 4 5 6 7 8 9 0   DOC/DOC   0 7 6 5 4 3 2 1

ISBN 0-07-137411-6

*The sponsoring editor for this book was Shelley Ingram Carr, the editing supervi-
sor was Stephen M. Smith, and the production supervisor was Sherri Souffrance.
It was set in Garamond following the TAB3A design by Victoria Khavkina of
McGraw-Hill's Hightstown, N.J., Professional Book Group composition unit.*

Printed and bound by R. R. Donnelley & Sons Company.

McGraw-Hill books are available at special quantity discounts to use as premi-
ums and sales promotions, or for use in corporate training programs. For more
information, please write to the Director of Special Sales, Professional Publishing, McGraw-Hill, Two Penn Plaza, New York, NY 10121-2298. Or contact
your local bookstore.

This book is printed on recycled, acid-free paper containing
a minimum of 50% recycled, de-inked fiber.

*To Paul "P.J." Weaver, USAF Academy,
Class of 1979: a man with a gigantic heart
and a zest for people, flying, and his country.
His life on Earth was snuffed out aboard his
C-130 by Iraqi ground fire during the
Persian Gulf War.
I miss you, P.J.; see you in Heaven.*

Series Introduction

The Human Condition

The Roman philosopher Cicero may have been the first to record the much-quoted phrase "to err is human." Since that time, for nearly 2000 years, the malady of human error has played out in triumph and tragedy. It has been the subject of countless doctoral dissertations, books, and, more recently, television documentaries such as "History's Greatest Military Blunders." Aviation is not exempt from this scrutiny, as evidenced by the excellent Learning Channel documentary "Blame the Pilot" or the NOVA special "Why Planes Crash," featuring John Nance. Indeed, error is so prevalent throughout history that our flaws have become associated with our very being, hence the phrase *the human condition*.

The Purpose of This Series

Simply stated, the purpose of the Controlling Pilot Error series is to address the so-called human condition, improve performance in aviation, and, in so doing, save a few lives. It is not our intent to rehash the work of over a millennia of expert and amateur opinions but rather to *apply* some of the more important and insightful theoretical perspectives to the life and death arena of manned flight. To the best of my knowledge, no effort of this magnitude has ever been attempted in aviation, or anywhere else for that matter. What follows is an extraordinary combination of why, what, and how to avoid and control error in aviation.

Because most pilots are practical people at heart—many of whom like to spin a yarn over a cold lager—we will apply this wisdom to the daily flight environment, using a case study approach. The vast majority of the case studies you will read are taken directly from aviators who have made mistakes (or have been victimized by the mistakes of others) and survived to tell about it. Further to their credit, they have reported these events via the anonymous Aviation Safety Reporting System (ASRS), an outstanding program that provides a wealth of extremely useful and *usable* data to those who seek to make the skies a safer place.

A Brief Word about the ASRS

The ASRS was established in 1975 under a Memorandum of Agreement between the Federal Aviation Administration (FAA) and the National Aeronautics and Space Administration (NASA). According to the official ASRS web site, *http://asrs.arc.nasa.gov*

> The ASRS collects, analyzes, and responds to voluntarily submitted aviation safety incident reports in order to lessen the likelihood of aviation accidents. ASRS data are used to:
>
> • Identify deficiencies and discrepancies in the National Aviation System (NAS) so that these can be remedied by appropriate authorities.

- Support policy formulation and planning for, and improvements to, the NAS.
- Strengthen the foundation of aviation human factors safety research. This is particularly important since it is generally conceded *that over two-thirds of all aviation accidents and incidents have their roots in human performance errors* (emphasis added).

Certain types of analyses have already been done to the ASRS data to produce "data sets," or prepackaged groups of reports that have been screened "for the relevance to the topic description" (ASRS web site). These data sets serve as the foundation of our Controlling Pilot Error project. The data come *from* practitioners and are *for* practitioners.

The Great Debate

The title for this series was selected after much discussion and considerable debate. This is because many aviation professionals disagree about what should be done about the problem of pilot error. The debate is basically three sided. On one side are those who say we should seek any and all available means to *eliminate* human error from the cockpit. This effort takes on two forms. The first approach, backed by considerable capitalistic enthusiasm, is to automate human error out of the system. Literally billions of dollars are spent on so-called human-aiding technologies, high-tech systems such as the Ground Proximity Warning System (GPWS) and the Traffic Alert and Collision Avoidance System (TCAS). Although these systems have undoubtedly made the skies safer, some argue that they have made the pilot more complacent and dependent on the automation, creating an entirely new set of pilot errors. Already the automation enthusiasts are seeking robotic answers for this new challenge. Not surprisingly, many pilot trainers see the problem from a slightly different angle.

Another branch on the "eliminate error" side of the debate argues for higher training and education standards, more accountability, and better screening. This group (of which I count myself a member) argues that some industries (but not yet ours) simply don't make serious errors, or at least the errors are so infrequent that they are statistically nonexistent. This group asks, "How many errors should we allow those who handle nuclear weapons or highly dangerous viruses like Ebola or anthrax?" The group cites research on high-reliability organizations (HROs) and believes that aviation needs to be molded into the HRO mentality. (For more on high-reliability organizations, see *Culture, Environment, and CRM* in this series.) As you might expect, many status quo aviators don't warm quickly to these ideas for more education, training, and accountability—and point to their excellent safety records to say such efforts are not needed. They recommend a different approach, one where no one is really at fault.

On the far opposite side of the debate lie those who argue for "blameless cultures" and "error-tolerant systems." This group agrees with Cicero that "to err is human" and advocates "error-management," a concept that prepares pilots to recognize and "trap" error before it can build upon itself into a mishap chain of events. The group feels that training should be focused on primarily error mitigation rather than (or, in some cases, in addition to) error prevention.

Falling somewhere between these two extremes are two less-radical but still opposing ideas. The first approach is designed to prevent a recurring error. It goes something like this: "Pilot X did this or that and it led to a mishap, so don't do what Pilot X did." Regulators are particularly fond of this approach, and they attempt to regulate the last mishap out of future existence. These so-called rules written in blood provide the traditionalist with plenty of training materials and even come with ready-made case studies—the mishap that precipitated the rule.

Opponents to this "last mishap" philosophy argue for a more positive approach, one where we educate and train *toward* a complete set of known and valid competencies (positive behaviors) instead of seeking to eliminate negative behaviors. This group argues that the professional airmanship potential of the vast majority of our aviators is seldom approached—let alone realized. This was the subject of an earlier McGraw-Hill release, *Redefining Airmanship.*[1]

Who's Right? Who's Wrong? Who Cares?

It's not about *who's* right, but rather *what's* right. Taking the philosophy that there is value in all sides of a debate, the Controlling Pilot Error series is the first truly comprehensive approach to pilot error. By taking a unique "before-during-after" approach and using modern-era case studies, 10 authors—each an expert in the subject at hand—methodically attack the problem of pilot error from several angles. First, they focus on error prevention by taking a case study and showing how preemptive education and training, applied to planning and execution, could have avoided the error entirely. Second, the authors apply error management principles to the case study to show how a mistake could have been (or was) mitigated after it was made. Finally, the case study participants are treated to a thorough "debrief," where alternatives are discussed to prevent a reoccurrence of the error. By analyzing the conditions before, during, and after each case study, we hope to combine the best of all areas of the error-prevention debate.

A Word on Authors and Format

Topics and authors for this series were carefully analyzed and hand-picked. As mentioned earlier, the topics were taken from preculled data sets and selected for their relevance by NASA-Ames scientists. The authors were chosen for their interest and expertise in the given topic area. Some are experienced authors and researchers, but, more importantly, *all* are highly experienced in the aviation field about which they are writing. In a word, they are practitioners and have "been there and done that" as it relates to their particular topic.

In many cases, the authors have chosen to expand on the ASRS reports with case studies from a variety of sources, including their own experience. Although Controlling Pilot Error is designed as a comprehensive series, the reader should not expect complete uniformity of format or analytical approach. Each author has brought his own unique style and strengths to bear on the problem at hand. For this reason, each volume in the series can be used as a stand-alone reference or as a part of a complete library of common pilot error materials.

Although there are nearly as many ways to view pilot error as there are to make them, all authors were familiarized with what I personally believe should be the industry standard for the analysis of human error in aviation. The Human Factors Analysis and Classification System (HFACS) builds upon the groundbreaking and seminal work of James Reason to identify and organize human error into distinct and extremely useful subcategories. Scott Shappell and Doug Wiegmann completed the picture of error and error resistance by identifying common fail points in organizations and individuals. The following overview of this outstanding guide[2] to understanding pilot error is adapted from a United States Navy mishap investigation presentation.

Simply writing off aviation mishaps to "aircrew error" is a simplistic, if not naive, approach to mishap causation. After all, it is well established that mishaps cannot be attributed to a single cause, or in most instances, even a single individual. Rather, accidents are the end result of a myriad of latent and active failures, only the last of which are the unsafe acts of the aircrew.

As described by Reason,[3] active failures are the actions or inactions of operators that are believed to cause the accident. Traditionally referred to as "pilot error," they are the last "unsafe acts" committed by aircrew, often with immediate and tragic consequences. For example, forgetting to lower the landing gear before touch down or hotdogging through a box canyon will yield relatively immediate, and potentially grave, consequences.

In contrast, latent failures are errors committed by individuals within the supervisory chain of command that effect the tragic sequence of events characteristic of an accident. For example, it is not difficult to understand how tasking aviators at the expense of quality crew rest can lead to fatigue and ultimately errors (active failures) in the cockpit. Viewed from this perspective then, the unsafe acts of aircrew are the end result of a long chain of causes whose roots originate in other parts (often the upper echelons) of the organization. The problem is that these latent failures may lie dormant or undetected for hours, days, weeks, or longer until one day they bite the unsuspecting aircrew....

What makes [Reason's] "Swiss Cheese" model particularly useful in any investigation of pilot error is that it forces investigators to address latent failures within the causal sequence of events as well. For instance, latent failures such

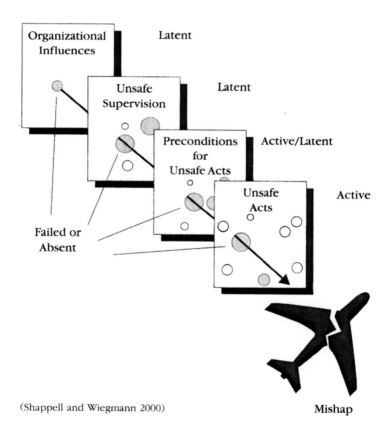

(Shappell and Wiegmann 2000)

as fatigue, complacency, illness, and the loss of situational awareness all effect performance but can be overlooked by investigators with even the best of intentions. These particular latent failures are described within the context of the "Swiss Cheese" model as preconditions for unsafe acts. Likewise, unsafe supervisory practices can promote unsafe conditions within operators and ultimately unsafe acts will occur. Regardless, whenever a mishap does occur, the crew naturally bears a great deal of the responsibility and must be held accountable. However, in many instances, the latent failures at the supervisory level were equally, if not more, responsible for the mishap. In a sense, the crew was set up for failure....

But the "Swiss Cheese" model doesn't stop at the supervisory levels either; the organization itself can impact performance at all levels. For instance, in times of fiscal austerity funding is often cut, and as a result, training and flight time are curtailed. Supervisors are therefore left with tasking "non-proficient" aviators with sometimes-complex missions. Not surprisingly, causal factors such as task saturation and the loss of situational awareness will begin to appear and consequently performance in the cockpit will suffer. As such, causal factors at all levels must be addressed if any mishap investigation and prevention system is going to work.[4]

The HFACS serves as a reference for error interpretation throughout this series, and we gratefully acknowledge the works of Drs. Reason, Shappell, and Wiegmann in this effort.

No Time to Lose

So let us begin a journey together toward greater knowledge, improved awareness, and safer skies. Pick up any volume in this series and begin the process of self-analysis that is required for significant personal or organizational change. The complexity of the aviation environment demands a foundation of solid airmanship and a healthy, positive approach to combating pilot error. We believe this series will help you on this quest.

References

1. Kern, Tony, *Redefining Airmanship,* McGraw-Hill, New York, 1997.

2. Shappell, S. A., and Wiegmann, D. A., *The Human Factors Analysis and Classification System—HFACS,* DOT/FAA/AM-00/7, February 2000.

3. Reason, J. T., *Human Error,* Cambridge University Press, Cambridge, England, 1990.

4. U.S. Navy, *A Human Error Approach to Accident Investigation,* OPNAV 3750.6R, Appendix O, 2000.

Tony Kern

Foreword

William Jennings Bryan once said, "Destiny is not a matter of chance. It is a matter of choice." Although I'm certain he was not talking about controlled flight into terrain mishaps, his words certainly fit. In modern aviation, all too often, a pilot unintentionally chooses the path that leads him or her to peril.

Nothing seems like a more unfitting end for the pilot personality than controlled flight into terrain. Think about it with me for a moment. Pilots are drawn to aviation for many reasons, but one of the most prevalent is the desire to control our own destiny. Many of us are drawn to flight because the sky is the one place where we can rely on our own preparation to see us through. When the wheels come up, we are in charge.

Self-reliance and self-determination come with a price tag, however. We will be held accountable for our actions, plans, and preparation—or the tragic lack thereof. Controlled flight into terrain is the ultimate in Draconian self-critique—and it is a humbling epitaph.

When it occurs we are all left wondering—how could it have happened? The NTSB sifts through the rubble in an attempt to figure out the sequence of events, and we mourn the loss of another friend. This scenario plays out repetitively over the world, and the purpose of this book is to assist you in making sure you aren't one of the unfortunate ones.

Controlled flight into/toward terrain (CFIT/CFTT) is, almost by definition, preceded by a loss of situation awareness. Lost SA can have many less severe manifestations, but when it is coupled with other factors, such as time compression, low fuel, or unfamiliar locations, the results are lethal.

As you will shortly discover, CFIT mishaps occur across the spectrum in aviation and strike the experienced and the inexperienced alike. They occur in the most sophisticated jet airliners armed with Ground Proximity Warning Systems, as well as the barest-equipped Piper Cub. No one seems to be immune. The causal factors are widely varied, but the results are tragically uniform. Almost no one survives a CFIT mishap, and few see it coming until it is too late.

Although a great deal of emphasis has been placed on technological solutions to the CFIT problem (as well as a good bit of human factors training on the subject for major airline flight crews), very little has been done to inoculate us from the CFIT threat. This is an unfortunate oversight and one that this book seeks to rectify.

There seem to be as many solutions to the CFIT challenge as there are causes, and this book takes on the toughest cases. After a brief introduction to the problems associated with CFIT, Dr. Daryl Smith, Lt. Col. USAF, looks at topics as widely varied as automation, time and planning, instrument procedures, weather, situation awareness, and complacency. Using real world case studies and personal insights, the text brings us face to face with impending disaster—but in the safe comfort of our favorite reading chair at ground speed zero. It is here where we avoid a future mishap, through contemplation and reflection on our own skills, equipment, and pilot tendencies. Through guided self-assessment, we can make positive changes to our techniques, and increase knowledge and proficiency in areas that will prepare us for the unseen traps in our aviation future.

Because CFIT/CFTT is the result of such a wide variety of inputs, developing a framework for understanding and providing error management techniques is a considerable challenge. It is our good fortune to be guided on this endeavor by Dr. Smith, a pilot and aviation psychology professor at the United States Air Force Academy. Daryl brings considerable practical and theoretical experience to bear on this complex issue. He has flown multiengine crewed aircraft and trainers. He has taught a variety of performance-related aviation courses, training the next generation of warrior pilots at the USAF Academy.

It has been my honor and pleasure to work closely with Daryl Smith for the past several years while stationed together in Colorado Springs. I know him to be one of the most sincere and dedicated aviation professionals in the world (as well as having a pretty decent turnaround jump shot in lunch-time basketball), thoroughly devoted to the cause of improving safety through pilot performance enhancement. Whether you are a paid professional pilot or an unpaid professional pilot (formerly known as an amateur), you will be a safer and more effective aviator for having read this book. A tip of the hat to Dr. Smith's first book-length effort. I hope it is not his last.

Tony Kern

Acknowledgments

Since this is a book about flying, I think it would be appropriate to thank those who helped me learn to "slip the surly bonds of earth...": Bob McDonald, Captain Wolfe, Jeff Wesley, Dave Pfeiffer, Ron Wanhanen, Dave Gladman, Lowell Stockman, Mark Ioga, Dale Kucinski, Greg Manglicmot, Mike Swigert, and Kevin Brox.

Thanks to my parents for their love and nurturing guidance as I was growing up. They were always encouraging me to pursue good things. They are very near the top of the greatest gifts I have been given in life.

Thanks to those who helped so much with this book: Bob "Huge" Chapman, who covered every nook and cranny of the training center; Kurt Driskill, my favorite ACE buddy, who got me hooked up with the fixes; and Gary, Doug, and Rod, my unexpected Delta guests, who gave me extra excitement for this project through their war stories.

Thanks to Bill and Lisa "Cold Fingers Super Mom" Stroble for clutch help in preparing the manuscript; and to Chris Wickens, who I have had the privilege of having as an instructor and a colleague—his writings and teaching have taught me a lot about aviation human factors.

Thanks to my wife Laura, the best wife on the planet, and my children Ryan, Stephanie, and Andrew, who bring so much joy to me. They stuck by me on a project that was a little bigger than I expected. My thoughts now return more exclusively to you.

Finally, I want to thank the Lord for giving me the grace and strength to persevere through this book. I am thankful for the lessons I learned during this project, lessons of what really matters in life.

Daryl R. Smith, Ph.D.

Controlled Flight into Terrain (CFIT/CFTT)

1

Introduction
to CFIT

The smokin' hole: a small room on the plains of West Texas where the young flyers would gather to discuss the week's flying war stories. They spoke of cheating death and punching holes in the sky. It was a small corner of the Reese AFB Officers Club, their Pancho Barnes Flying Bar on the edge of their Edwards AFB Murdoch Dry Lake bed. Many a successful checkride was celebrated there. On Friday afternoons, the young bachelor student pilots would gather around the free hors d'oevres, that night's dinner, and share a beverage as they discussed what they learned and hoped to learn.

A smokin' hole: what none of us wants to become. A little part of us shutters when we see or hear of one. The thought leaves you wondering, a bit shook. In the back of your mind you realize "it could have been me." At least for any pilot who has flown for a while, the realization comes that on another day, in another time, or in another place that could have been him or her. There are those pilots like Chuck Yeager, Pete Conrad, Wally Schirra, Michael Collins, the young Navy and AF test pilots in the 1950s portrayed in *The Right Stuff* who won't allow this to be verbalized. It's against the culture. It eats into a young aviator's confidence, but the wives and the husbands know the reality in the back of their minds—it could have been me. The real young pilots are still perhaps a bit naive; it *can't* happen to them. The older pilots realize, there but for the grace of God go I.

What I desire to re-create here is *the* smokin' hole, a place where we can gather and learn from our war stories. As Captain Fisher used to tell me on those plains of West Texas, "When you stop learning, you start dying." That scene at Reese AFB has been re-created a thousand times over around flight desks and break rooms at FBOs, base ops, and O clubs around the world. This book is an attempt to emulate that process, but in a more systematic fashion. We must continue to learn or we begin to die.

What's the Problem, Mac?

Remember those old 1950s black-and-white pictures? Often there would be some guy standing there staring, kind of dumbfounded, when some blue-collar guy would walk up and say, "What's the problem, Mac?" or "Oh hey, buddy, you got a problem?". Then the dumbfounded guy would utter something like "No, no, there's no problem," and then he would wander off kind of aimlessly while the eerie music started. The guy was off to either confront the problem that didn't exist or ignore the problem and pretend it would go away. "I must have been seeing things." Guys are good at denial.

Most people won't take action until there is a problem. Some will still take no action unless they perceive potentially serious and/or personal harm. Some just never act (those are the walking comas of the world). How do they get through life?

Those who have flown for a while have all had at least one "near miss." Those that are relative novices may still be waiting for theirs. But like the old failed checkride adage goes, "there are those that have and those that will." A near miss can be another aircraft, something on the ground, or the ground itself. It gets your attention real fast. Afterward there is a numbness, then the awful realization of what could have happened. It gets you focused in a hurry.

The Flight Safety Foundation published an analysis of controlled flight into terrain (CFIT) accidents of commercial operators from 1988 through 1994. They examined 156 accidents. These accidents alone accounted for 3177 deaths. In 97 percent of the accidents where the data were known, the aircraft was completely destroyed. CFIT isn't usually "just a ding," it is usually fatal.

For a few pilots every year, the near miss becomes a "didn't miss." Worldwide from 1988 through 1994 there were an average of 18 CFIT accidents per year, but that is just among major, regional, and air taxi operations. That does not count general aviation or military accidents. Think of the Rose Bowl full of spectators. You and I, the pilots, are those spectators. Twenty of us could go from near misses to didn't misses this year. That's like a whole seating section of 20 people—gone. You say, "20, that's not a lot of people." It is if it's your section.

Without proper action, one of those didn't misses could be you. Controlled flight toward terrain (CFTT) becomes controlled flight into terrain—a big difference. That is a life-losing difference. So what's the problem, Mac? CFIT, that's the problem. It's my problem and your problem. So let's figure out the solution together. Don't be a "didn't miss."

A Picture Is Worth a Thousand Words

Every great teacher knows implicitly the power of a story to teach. Good stories are intrinsically interesting and provide lessons and principles to emulate or avoid. Great teachers throughout history have used them...Jesus Christ and his parables, Aesop and his fables. Stories and pictures are what we will use here because they work.

In these pages the stories are case studies. A case study is simply a well-documented series of events put together by an investigator, designed to inform or enlighten the reader of some important principle(s) or facts. Often these case studies chronicle events that are rare and unique. These events would be difficult or unethical to reproduce in a laboratory for scientific investigation. Often a case study provides critical information on an unclear and/or high-stakes phenomenon. Our goal here is to learn all we can from these stories, to glean their principles in order to avoid the harmful high-stakes consequences. That is our purpose. But before we proceed further, I must be clear about what we are discussing.

The Orchestra Is Warming Up

Ever hear an orchestra warm up? Sounds horrible. But a well-tuned orchestra playing the same song is a piece of art. Imagine paying $30 a ticket just to hear an orchestra warm up. Ridiculous, isn't it? But the point is this: We need to get on the same sheet of music; to do that we must define some terms. This book is about controlled flight into terrain. According to the Flight Safety Foundation, CFIT is an accident in which an otherwise serviceable aircraft, under the control of the crew, is flown (unintentionally) into terrain, obstacles, or water, with no prior awareness on the part of the crew of the impending collision. It is flying a perfectly good aircraft into the ground, and usually the pilot has no clue of what he or she is about to do.

A key point is that the aircraft is not in an uncontrollable (e.g., a spin) condition when it hits the ground. An example of CFIT would be when a pilot under visual flight rules (VFR) enters a cloud deck and slams into a mountain. The pilot had control of the aircraft upon impact. A counterexample is the John Kennedy, Jr., crash in the summer of 1999. Kennedy entered instrument meteorological conditions (IMC) when he was supposed to remain in visual meteorological conditions (VMC), and, as a result, he induced a vestibular illusion, lost control of his small craft, and plunged into the sea. That is not what this book is about. Such an accident is covered amply by another book on spatial disorientation elsewhere in this series. Both of these accident involve weather, but in one the aircraft is under control and in the other it is not.

I will also cover controlled flight toward terrain: flying at an altitude and attitude that would, if continued, result in contact with terrain. Let's return to the previous example. Eagle 1 is under VFR, yet enters a cloud deck. When the pilot pops out, he sees mountainous terrain rising rapidly in front of him. He is able to take evasive action moments before possible impact and survives. Had he not taken some form of action, CFIT would have occurred, but this is CFTT. I will examine the slightly more positive CFTT as well as the more negative CFIT, as we tend too easily to fixate on the negative. We should strive to learn from CFTT too. One important note: We read first-hand accounts of CFTT; we generally can't read first-hand accounts of CFIT.

None of the "first-handers" are usually around to do the writing.

Beware the Monday Morning Quarterback

There is a danger when using CFIT case studies. The danger is developing a sense of arrogance or a tendency to become judgmental. How could that pilot have been so stupid? I would never do that. How did he miss it? And on and on.

A well-documented psychological phenomenon is called the *hindsight bias.* Hindsight bias is the tendency to believe, once an outcome has occurred, that we knew all along how it would turn out. Knowing what happened makes that outcome seem to have been more likely in the first place (i.e., in hindsight), compared to how likely it actually seemed before it occurred (i.e., in foresight). After the event, we seem to automatically assimilate new and old information. After the outcome, people generate a plausible explanation of how they could have predicted the event beforehand (i.e., they assimilate the old knowledge to the new), making the actual outcome seem more inevitable within the reinterpreted situation. Thus, people recall the past in ways that confirm the present. For further explanation of this phenomenon, see the For Further Study section at the end of the chapter.

To you and me this is called being a Monday-morning quarterback. Things are always much clearer on Monday morning than Sunday afternoon. Why did the coach call that play? Why did he throw that pass? And on and on. Hindsight is 20/20, or better. I will strive to avoid that Monday-morning-quarterback mentality. I abhor arrogance. Our goal is to learn and to live—not to judge those who have fallen before us.

The Route of Flight

Every good flight should begin with an initial, well-planned flight route. If you have no goal, you will hit it every time. Each chapter will contain a series of case studies. These studies are taken from actual Aviation Safety Reporting System (ASRS) incident reports, National Transportation Safety Board (NTSB) accident reports, and personal experience or interviews. Each case study strives to relay the basic facts as accurately as possible. Any gaps in the accident narrative I will fill in using literary license. This license, however, will in no way alter the basic facts of the case. The license is used only to bring the facts to life.

I said earlier that I want us to learn from each case study. That will occur using the three elements that earmark this series of books on controlling pilot error. The first element to "controlling" error is *preparation*. What knowledge, skills, and attitudes are necessary for an individual to effectively avoid these errors altogether?

Second, if the pilot encounters this error or challenge *in-flight*, what can be done to mitigate it—to keep it from becoming an error chain? This has been referred to as *error management* and involves real-time judgment and decision making to keep a situation from deteriorating.

Finally, I will advocate and address a *post-flight analysis* or lessons-learned approach with each case study to help us understand that errors are learning experiences, and the real key is to prevent their reoccurrence.

Consistent themes emerged from an analysis of the ASRS reports. These themes form the basis for the following chapter subjects: weather; VFR to IFR (instrument flight rules), situational awareness and complacency, crew resource management (CRM), automation, equipment problems, controller error, problems with instrument procedures, planning and time, and fatigue and other factors. These chapters have a large degree of overlap with factors associated with CFIT as well as the CFIT taxonomy as identified in the Flight Safety Foundation report. Each chapter will be illustrated by case studies (both CFTT and CFIT), with a focus on the chapter subject at hand but including ties between chapters.

Chapter 2 will investigate how the forces of Mother Nature have a huge impact on CFIT. We will investigate the reality of having to turn around and why that is so difficult to do. The importance of knowing your aircraft limits and your equipment will be discussed. It is difficult to fly into unfamiliar terrain, and doing so requires that you use all the resources available to you as a pilot. Pilots are often tempted to become too focused on one aspect of the mission during bad weather. We will discuss this phenomenon at length. Chapter 3, on VFR to IFR, is a special subset of weather problems. This has been a particularly lethal problem for general aviation pilots, though others are not immune.

Lost situational awareness was responsible for 45 percent of the CFIT incidents in the Flight Safety Foundation report. There are several factors that lead to lost situational awareness (SA) and these will be focused on in Chap. 4. One of the best ways to avoid lost SA is through good crew coordination; Chap. 5 introduces the concept of CRM. Some of the most famous CFIT cases that we cover will concern poor CRM.

One of the biggest challenges of CRM is to use automation appropriately. Three major automation innovations have occurred in the last 25 years, and we will discuss these in Chap. 6. Automation has brought blessings as well as baggage. One of the innovations, the Ground Proximity Warning System (GPWS), has been responsible for a reduction in CFIT over the last decade. Seventy-five percent of the mishap aircraft in the Flight Safety Foundation report lacked a GPWS. Proper use and abuse of the GPWS will be discussed. The automation discussion will lead into other types of equipment problems and their role in CFIT, covered in Chap. 7. Automation machinery is not the only type of equipment relevant to this topic.

To err is human, and air traffic controllers are no exception. In Chap. 8 we will examine when and how their errors are most likely to occur. That will lead into a discussion in Chap. 9 of pilots practicing poor instrument procedures as a causal factor in both CFIT and CFTT.

A recurring theme throughout this book is the role of the concepts of time and planning in CFIT. This theme is examined in depth in Chap. 10 and the two concepts will be shown to be closely related. They play on one another and are influenced by fatigue and other factors as well, which will be discussed in Chap. 11. The discussion of factors is not exclusive to pilots and controllers, but has many applications outside of aviation; we will also examine the role of the organization in CFIT.

The academic realms of human factors engineering and aviation psychology have much to teach us on the subject of CFIT/CFTT. I will generally translate the more academic jargon into layperson's terms when using or applying these principles. The hindsight bias/Monday-morning quarterback mentioned previously is a good example of this practice.

Though my background is in military aviation, I use a variety of examples throughout this book. Case studies will originate from the major airlines, the regional airlines, the air taxi sector, general aviation pilots, and the military. I think you will find that there is something for everyone. We all can learn from all types of aviation when it comes to CFIT.

My goal here is a readable style. There are more academic approaches to this subject found in the scholarly journal literature, and if that's what you want, you'll have to look elsewhere. The book is written for the common aviator, and the academic is invited to learn as well. I envision every pilot, regardless of experience, from a person taking flying lessons to the most senior airline captain, taking something useful from this work.

Which Hole Will It Be?

As you read this book, you have a decision to make, as do I. We have the choice to learn and to live, or we can stop learning and start dying. If we choose to learn, we can then apply what we've learned the next time we step to the plane. Let's choose to learn at *the* smokin' hole, so we don't become *a* smokin' hole.

References and For Further Study

Bryant, F. B. and J. H. Brockway. 1997. Hindsight bias in reaction to the verdict in the O. J. Simpson criminal trial. *Basic and Applied Social Psychology*, 7:225–241.

Khatwa, R. and A. L. C. Roelen. April–May 1996. An analysis of controlled-flight-into-terrain (CFIT) accidents of commercial operators, 1988 through 1994. *Flight Safety Digest*, 1–45.

2

Weather

My first real appreciation for the weather's role in aviation came as a young cadet standing on the terrazzo of the U.S. Air Force Academy. The terrazzo is a huge plaza where the cadets form up into squadrons in order to march into Mitchell Hall for lunch. As those with military experience can attest, these formations can become monotonous and routine, and noon meal formation is no exception. One day while standing in formation, I took a peek skyward, where I spotted a glider nearly over head. A few minutes later another quick peek revealed another glider in the same location. With another glance (cadets are supposed to have the eyes steely caged on the horizon) moments later to the same spot, the glance became a gaze. It was the same glider, but it was not moving over the ground. With a normal speed of 55 KIAS (knots indicated airspeed) and a 55-knot headwind, the glider had a ground speed of zero. Thus, the glider appeared not to be moving. I was amazed.

The Power of the Wind

Later in the T-37 and T-38 jet aircraft over the west plains of Texas, I got another lesson in appreciation for the weather. The slower and squattier jet, the T-37 was able to withstand a much greater crosswind than the sleek supersonic T-38. It was a case where the tortoise often flew as the hare sat by. I was amazed as the 2000-lb aircraft crabbed into the wind immediately after takeoff. I didn't like it and immediately put in crosswind controls to straighten the bird. The IP (instructor pilot) said, "No. Let her crab." I was amazed at the power of the wind.

The real eye-opener came as I transitioned to the large KC-135 refueling aircraft. Again, the crab was there immediately after takeoff. But what really got my attention was the aerial refueling. We would refuel fighters, which was fun, but the "heavies" such as a C-5 and B-52 were the showstoppers. These huge aircraft came in 40 ft in trail. I remember the first time I saw a B-52 (Buff) in trail when we hit some light chop. The wings of the Buff would literally flap up and down (roughly 3 ft) like a big bird. The same thing happens with the C-141s and C-5s. An amazing sight, all that heavy metal moving, flapping, and twisting around in the air. A man could never move those wings around like that. The weather made it look easy. Once again, I was amazed at the wind's power.

How serious a problem is the weather? The NTSB studied accident files over a nine-year period. In that period, 4714 people died in 2026 accidents. Weather-related accidents accounted for more than one out of three of all fatal general aviation accidents (Craig 1992).

The buzz phrase in the 1960s in military flying was "all-weather capability." All-weather maybe—if the conditions were just right. All-weather fighters, such as the F-106 Delta Dart, were all the rage. We have come a long way toward all-weather capability, but it's still all-weather if all the conditions are right (not too much wind, rain, shear, snow, ice, and so on). With this increased capability comes an increased responsibility to plan and fly wisely in and around the weather. The power of the weather is amazing. Mother Nature doesn't play games. It is to this end that we turn our attention now—you, aviation, and the weather.

The Iceman Cometh

Know your aircraft limits and know them cold—no pun intended. In the T-37 we had a little memory device: "Race through trace, climb through rime." This simply meant that the T-37 was capable of cruise through trace icing conditions. The Tweet could climb through rime icing conditions.

Remember, icing can significantly decrease aircraft effectiveness. It does so by increasing drag, decreasing lift, and thereby decreasing performance. Icing forms around the airfoil, disturbing the smooth flow of air over the wing. Several aircraft (especially larger aircraft) have anti-icing devices on the wings—specifically, the leading edge of the wing. Though it is mostly standard equipment, one should not forget the pitot tube heat anti-icing device as well. Many a pilot turns on the pitot heat even in clear weather as a preventative measure and a good habit pattern. But some tubes can overheat in warm weather and reduce pitot tube life.

The subject of weather is covered by another book in the Controlling Pilot Error series. The icing information is meant to illustrate the point that we as pilots have an obligation to know our aircraft, our environment, and our limitations. This knowledge can help us avoid the types of incidents we will examine next.

I Want That Mountain

"My name is Chip and I was on a night cross-country flight from Henderson to Sacramento. I am an instructor pilot in the Skyhawk 172. As the flight progressed, I encountered some snow and trace icing, though I remained VMC. After I finished speaking to Reno (RNO) radio around HIDEN intersection on V135, I decided to turn back to Henderson; it was too dicey ahead. During the return trip, I began to daydream and to come up with alternative plans now that my trip to Sacramento had been nixed. Next thing I know, I became disoriented as to my position. My DME wasn't functioning properly, and navigation radios were also squirrelly. The plan was to fly south, until clearing Mount Charleston, then turn east. That was the plan. Instead, I very nearly hit Mount Charleston, which frazzled my nerves a bit! I could see the lights of Las Vegas around the mountain, and without thinking, I turned toward Vegas. Las Vegas Approach cleared me through the class B airspace back to Henderson. They didn't say anything about my position, which I noticed only later was within the A481 restricted airspace. After my close call with Mount Charleston, and my eager desire to return home, I didn't think to double-check my charts for restricted airspace (and Nevada has a lot of it). I just saw the lights and turned in to them, thinking only of the class B airspace. A major contributing factor was that in my haste to leave, I hadn't bothered calling for a weather briefing. Had I obtained one, I never would have gone in the first place!" (ASRS 419796)

Forecast—too secure

It goes without saying that a weather forecast prior to takeoff is a necessity for safe flight. This is a no-brainer, yet sometimes pilots skip it, even instructor pilots such as Chip. Sometimes experience can cause us to let down our guard. We get a false sense of security. Some say that the most dangerous pilots are new pilots and really experienced instructor pilots. There is a famous case in the Air Force of a KC-135 running out of fuel in Michigan with a Standardization and Evaluation (Stan Eval) pilot at the controls. The Stan Eval pilot is supposed to be the most experienced and expert pilot on base. Yet he, not the young aircraft commander, made the decisions that led to the air refueling aircraft running out of fuel. Very experienced pilots can become complacent with accompanied lack of attention to detail. They become too comfortable with flying. I guarantee Chip knew better than to take off on a night cross-country flight without a weather forecast. Beware of experience!

An honest forecast

When it comes to forecasts, the first rule is to get one. The next rule is to get a good one. Don't play games. And games can be played. I heard once of military aircrews at a particular base that had trouble getting good forecasts (read: forecasts that would allow them to file a flight plan and be within regulations) from a particularly stingy forecaster. This old forecaster had established a reputation of not only leaning toward the conservative in forecasts but falling down in that region. If there was any possibility of icing conditions en route, he would put it down on the weather forecast sheet, however remote that chance. If there was question of trace versus light rime, he would always default to the worst of the two, light rime. Often crews could not fly for several hours because of these forecasts. Or they would wait until a less conservative forecaster came on duty. However, reportedly, sometimes instead of waiting, they would put down a cruise altitude on the weather briefing request that was below the minimum en route altitude (see Chap. 8 for specifics on this and other minimum altitudes) for that jet route or victor route. The altitude given was low enough that the chance of an icing forecast was greatly reduced. Once they received the "good" briefing (meaning no icing mentioned), they would actually file a flight plan at the correct minimum en route altitude (MEA) and be on their way.

Of course, here is the rub with this strategy. If you play this game, you fly at the MEA and you have clear sailing—for a while. Then you hit some icing more severe than your aircraft or regulations allow. Icing is also above you, so you have to descend. But you can't descend safely; you are at minimum en route altitude which allows you 1000-ft clearance (2000 ft in mountainous terrain) from the ground. If you are lucky, a controller could perhaps vector you at minimum vector altitude (MVA), which is *sometimes* lower than the MEA. If it is not, or you can't get a MVA vector for some reason—uh oh, you better hope you can turn around and get out of the icing as soon as possible or that you get through the icing quickly. You also better hope that any anti-icing equipment you may have is functioning properly, especially your pitot heat! So if your estimation is wrong in this game, "you are hosed," as the vernacular goes. All in all, "the good forecast game" is probably not a game that is worth playing.

Remember, even good forecasters are wrong sometimes. For example, in Seattle, Washington, a good forecaster is wrong maybe 15 percent of the time. Weather forecasting in some parts of the country is more accurate than in others. So realize that many forecasters give the conservative forecast for good reason: erring toward the safe side. Don't compromise the built-in safety factor without good reason. Also remember that predictability changes with the seasons. Weather in the fall and spring is often less predictable; forecasters are more often wrong at these time. Weather can easily be worse than expected. On a sunny morning in the fall, you could expect an Indian summer afternoon and not get it.

Home field advantage

There is one more strategy to keep in your hip pocket. After flying in your local area for a while, you become very familiar with the weather patterns. You can use this knowledge of your environment to your benefit. For example, during summertime in Colorado Springs, you should expect scattered thunderstorms late in the afternoon. Bank on it, even if they are not forecasted. This is an example of knowing your home area, which can be a great advantage. Of course, you don't have this luxury flying cross-country in unfamiliar territory, but maybe you can call an FBO and pick someone's brain.

Don't go there

Chip traversed restricted airspace and didn't realize it until later. It was almost undoubtedly "cold" (a "hot" area is an area currently under use, usually by military aircraft), since the controller did not mention the aircraft's errant position to the pilot. Restricted areas can be transversed with controller approval if they are not in use. However, you need to be aware. Restricted areas are used by aircraft that are frequently exceeding "normal" attitudes and bank angles. The restricted areas in Nevada can be used for mock dogfighting or tests of classified aerospace vehicles. Violating a restricted area can be hazardous to your flying career. Why would Chip fly through this area? What was he thinking or not thinking?

Your bags, sir

Often disappointment or a change of plans can bring dejection and distraction. It causes you to refocus, replan, and begin anew. You must decide on new goals and objectives. If you are not prepared with a backup plan, this distracting situation is compounded, as you must come up with something from scratch. Unfortunately, while a person is refocusing or ruminating over a new plan, attention to the tasks at hand suffers. According to Chris Wickens, cognitively, we only have two attentional resource pools. We have cognitive attention pools that can be directed at visual/spatial-type functions and another at phonetic/language-type functions. The pools are limited, and if we deplete them while focusing heavily on one task, we can find them empty when we try to direct them simultaneously to another.

In layperson's terms, we are talking about something called "emotional jet lag." Economists call it "sunk costs"; since the loss has already occurred, it does no good to continue throwing money at the problem. I think we have all experienced this phenomenon. When something happens to us, we get so focused on the event that just occurred that we fail to focus on what is at hand or just ahead. Or we can be so preoccupied with coming up with plan B that we fail to concentrate on what we should be doing. Many pilots have experienced this phenomenon on a checkride. They make a mistake in front of the evaluator, and they can't get over it. They let it eat at them until they are so distracted or disappointed that they make another (often more serious) mistake. The remedy here is to put the situation behind you and move on. Chip didn't move on, and he almost got "moved out"—by Mt. Charleston. Perfectionists are especially prone to emotional jet lag. If you are a perfectionist, realize you will have to fight hard against this one. It is also important to realize that emotional jet lag can lead to error chains, which are discussed next.

A multifaceted problem

The final lesson we can glean from Chip is that often there are many factors at play in a CFTT or CFIT incident. In Chip's case he ran into bad weather. He failed to get a weather briefing, so he was caught off guard. Because of this, he got distracted. Compound those factors with the fact that his instruments began to act up: "My DME [distance-measuring equipment] wasn't functioning properly, and navigation radios were also squirrelly." Talk about a case of bad timing. Bad weather is not the time you want to have your instruments become untrustworthy. It well could have been that the icing conditions contributed to the instrument malfunctions, in which case Chip had only Chip to blame. The moral of this story is that CFTT and CFIT usually have multiple factors, known as an *error chain*, leading up to the finale. This theme will be repeated throughout this book.

I mentioned earlier the home field advantage a pilot has when flying near his or her home base or airport. Flying into unfamiliar territory can bring some interesting challenges, as our next pilot finds.

Things That Go Thump in the Night

It is a dark and stormy night as we pick up this crew of a Boeing 727 heading toward unfamiliar territory carrying freight.

"We were preparing for the approach at Belize City just south of Mexico. I was the first officer and the pilot flying (PF). Small thunderstorms were in the area as we flew along in our Boeing 727. There was no moon, no approach lighting system, no VASI (visual approach slope indicator), and no precision glideslope available at Belize. Although the airport has PAPI, it was out of service (OTS). There were no surrounding lights to provide better depth perception and it was very dark. Winds were reported variable, from calm to 100 degrees at 18 knots gusting to 25 knots to 010 degrees (up to a 90-degree shift) at the same velocity. We were cleared for a VOR approach to Runway 07.

"At start of approach, winds were calm, then variable. At 7 mi inbound an area of rain became visible. We had good visibility until rain started falling heavily. We were crabbed to the right of Runway 07, but winds were reported 010 degrees at 18 knots gusting to 25 knots by tower (which would have required a crab to the left—things weren't adding up). We had the runway in sight with the VOR on. We left the minimum descent altitude of 440 ft at 1.5 DME (inside of 2 mi on the approach) and the heavy rain hit. The flight conditions were now turbulent and gusty. We were at 350 ft, field still in sight. Very suddenly, we were at 240 ft. It wasn't a drop, but more of riding a wave down. Then the aircraft sort of settled. We saw that we were low and pushed to go around thrust. Both I and the captain pushed the power up to max. We had been carrying 10 knots extra with the bad weather. The airplane responded and we were climbing.

"As the aircraft accelerated for a go-around, we felt a light impact and loud thump somewhere on the aircraft. The airport lighting was so poor at Belize that we decided not to make another approach, so we headed north, northwest and diverted to Merida, Mexico. Immediately after our landing and parking at the gate at Merida, we conducted a postflight inspection. During this inspection, we saw where the #8 leading-edge slat was dented from a tree strike. Furthermore, there were some tree branches stuck in the landing gear. We had clearly hit a tree with a flap. Fortunately, we were able to make that climb out and divert." (ASRS 420872)

Disclaimer

Let me say right up front that this incident may not be considered by some to be a CFIT incident in the purest sense of the term. The aircraft encountered wind shear, so it could be argued that it was not under controlled flight when the incident occurred. The flip side is that the aircrew was exiting the windshear effects and the aircraft was under control when the tree strike occurred. Furthermore, the aircrew made some questionable decisions while the aircraft was clearly under their control that led to the possibility of a strike. Finally, the ASRS reporting authorities chose to categorize this incident as a CFTT in the anomaly description area. For these reasons, and the fact that this case study presents a wealth of lessons, I have chosen to include the study.

Toto, we're not in Kansas anymore

As mentioned earlier, when we fly away from familiar territory, we lose the home field advantage. We can't tell for certain that this was new terrain for this aircrew, but with major carriers adding pilots monthly, we can be sure that this type of situation will present itself to new-comers at some point. In new territory, we won't be as familiar with local weather patterns and our destination may not have the creature comforts that we are used to at our home drome. This is compounded when flying outside of the United States, as the 727 freight carriers found. There was no approach lighting, no glideslope information, no VASI or PAPI available, and no moon to help light the way. This would be very rare at a major U.S. airport. So our crew was behind the power curve before they ever began the approach.

Let the wind blow

There were thunderstorms in the area, which should always alert the crew to the possibility of wind shear. Wind shear is a sudden change in wind direction or speed of such a magnitude that an aircraft's ground speed can be severely affected. Wind shear has been responsible for some major aircraft accidents in the previous decades. The crash of the Delta L10-11 at Dallas/Ft. Worth in August 1985 is one of the most notable. Beginning in the late 1980s, major U.S. airports began installing wind shear detectors. Denver, notorious for tricky winds, was the first to install the system. Devices are located across the field at the approach and departure ends of the runway. The sensors are interconnected to the terminal Doppler warning radar (TDWR). The radar is monitoring for wind deviations. The system is set to look for gradients and has parameters that indicate whether a deviation is categorized as wind shear or the more dangerous microburst. These detectors have helped to lower the incident of wind shear–related incidents.

It is likely that no such instrumentation exists at Belize City. Therefore, the crew was further behind the power curve because of a lack of automation. If the aircraft was equipped with color weather radar, the crew then had another tool at its disposable in the battle against the weather. Color radar can identify the most hazardous thunderstorm cells so that the aircrew is able to steer well clear of them. Of course, there is the old trick of looking for the anvil of the thunderstorm and going the other direction. The anvil is the portion of the cloud where the storm is spitting out its worse stuff. Of course, this crew did not have the anvil option available on a dark night.

Just hold on there

One option that this crew seems to have overlooked is the holding pattern. Spending 20 minutes in a holding pattern can allow conditions at the field to get much better. Crews too often ignore this option. It seems we too often want to press to get to our destination. This is known as "get-home-itis" or "get-there-itis." This phenomenon will be discussed in Chap. 3. From the report the pilot states that there were small thunderstorms in the area. With a series of small thunderstorms, it would be possible to simply wait until a small cell passed the airport. The crew noted the variable winds being called by Approach and Tower. Furthermore, it had direct evidence of the instability of the situation when Tower was calling winds that required a left crab, yet the crew needed a right crab to maintain course guidance. Assuming there was ample fuel on board and the weather was not forecasted to become even worse, it would have been wise for this crew to enter holding until the winds and weather stabilized.

The good stuff

To the crew's credit, it did some things right—things we can learn from. James Reason points out that in-flight mitigation of errors is key. This crew made an error that resulted in entering wind shear and a subsequent tree strike. But it didn't let the thump become emotional baggage. It moved on. Second, when entering variable wind conditions (and possible wind shear), the crew kept an extra 10 knots of airspeed. This is a good idea. I was always taught, if Tower is calling a wind gust, carry the number of knots extra on the approach to negate the gust. In this case, winds of 18 gusting to 25 knots were called. So the crew needed to carry at least 7 extra knots of airspeed. It elected to use 10 knots, which came in handy. It is likely that the crew's company (and many other organizations) has standard operating procedures (SOPs) concerning appropriate procedures under these circumstances. The extra 10 knots probably saved this aircrew.

The crew also makes a good call in immediately executing a go-around once it felt the plane settle or "ride the wave." Both the PF and the captain went for the power, and the PF immediately raised the nose. This is the correct response. You need to get away from the ground. Of course, you don't want to be too abrupt in pulling the nose up. If you do, you risk a stall, and then you are really in trouble. Again, most organizations have SOPs concerning this maneuver. The key is not to hesitate. Once wind shear occurs, react with power!

Finally, the 727 freight crew was very wise to divert to another airport with better conditions. It could have elected to remain at Belize, go into holding until things settled down, and then tried the approach again. But at this point, it was pretty shaken. In addition, the lighting and approach procedures available at Belize were not the best, and Merida is not too far away (about an hour) and is a larger facility. Bad weather, with poor lighting at night and no precision guidance available, made Belize a rather unattractive option after the strike. Of course, the crew risked that the thump it heard may have caused a structural or internal problem that may have been worsened with an hour's divert flight. But its first priority was to safely land, which is smart.

Epilogue

So what happened to this crew? According to the ASRS callback conversation with the reporting pilot (the PF), there were some repercussions. The flight crew underwent further training in a company simulator flying several nonprecision approaches. The FAA is taking further action by asking the company to move the captain to first officer and the first officer to second officer for a period of six months. It is unknown whether the company followed the FAA's recommendation.

Clipper by the Sea

Bob is a small-aircraft pilot with about 1000 hours who flies primarily for pleasure or recreation. He had about 750 h in the C177 he was flying the day of this incident. He hadn't flown a lot lately, around 15 h in the last three months. This is his story.

"On December 18, 1998, at approximately 2145, I requested taxi to Runway 34 at Bellingham (BLI), a couple hours' drive from Seattle, with the current ATIS [Air Terminal Information System] information. Since it was December, it was very dark by 2145 hours. During the call I stated that we were southwest bound (I had a passenger) but wished to go with the understanding that if the weather in that direction was not suitable, we would like to return. Our destination was Orcas Island airport (ORS). The tower controller informed us that the terminal information was in the process of changing and gave us a new briefing with 7-mi visibility, down from 10 mi. He asked our destination, and after I responded, he informed us that he had no current information about conditions in East Puget Sound, which was the Orcas Island area. Bellingham was at that time reporting a few clouds at 800 ft, 1400 ft overcast, light rain, wind 360 degrees at 13 knots.

"About 10 min earlier, I had observed a flight execute an ILS approach to Runway 16, overfly, circle west, and land on Runway 34, and I noticed no obscuration caused by low clouds. When I reached Runway 34, I was cleared for takeoff and began an uneventful takeoff.

During climb out at 500 ft MSL (mean sea level), I encountered clouds and began shutting off the landing/taxi lights and strobe to cut down on the reflections in the clouds. I radioed the tower and indicated that I was returning to land. The controller responded that he was increasing the landing light intensity to max (so that I could see the airport) and would amend the weather information immediately. I leveled out and began a 180-degree turn to the left.

"During this turn, I apparently overreacted and increased the bank angle to the extent that I began losing altitude. My instrument scan was poor, and I realize now that I was spending too much time looking outside for ground references. I had leveled the wings but was still descending when I saw treetops by the light of the navigation lights. I did an immediate pull up/power increase and we felt a shudder. At that time I felt that the left gear had impacted a treetop and that the only airworthiness concern would be upon landing. The runway was now visible off the left side and we were parallel on a close in downwind. I reported 'runway in sight on left downwind,' and the controller reduced the intensity of the runway lights and cleared me to land. He also informed me that he was amending the conditions to 300 ft broken and thanked me for the ceiling information, since his ceilometer was out of service.

"The landing was uneventful and the gear and braking system seemed to be normal. During taxi the controller mentioned that the left navigation light seemed to be out, and I informed him that I would check it. We taxied to transient parking, tied down, and deplaned. Upon inspection, the left wing leading edge had obviously impacted a vertical obstruction approximately 2 inches in diameter and had crumpled a portion of the skin adjacent to the taxi/taxi light lens, which was broken. Some damage to the navigation light was evident, but because of the heavy rain and a need to catch a ferry, we left very shortly after landing." (ASRS 423792)

A little rusty?

We can learn several things from Bob's aborted pleasure flight. The first thing to notice is that though Bob had 750 h in the aircraft, he didn't have a lot of hours as of late. He had roughly 15 h over the last 90 days. If we assume a 1.5-h total per sortie, that is 10 flights over a 90-day period, or one every nine days. (Of course, we are assuming the flights were spread evenly.) Many general aviation pilots find their flights come in spurts. He could have had three or four flights in the last two weeks, or they could have been 80 days ago. Whatever the exact spread, we have reason to wonder about Bob's proficiency, an issue we will return to later.

Bob, the man with the plan—the 5 Ps

Are you familiar with the 5 Ps? They are essential not only in flying, but in life. Bob practiced the 5 Ps and it came in handy. Notice Bob's words: "wished to go with the understanding that if the weather in that direction was not suitable, we would like to return." Bob had a plan and a backup plan before he took off. Bob was a man with a plan.

I remember listening to a Navy SEAL (Naval Special Forces) once speak about an operation during the Gulf War off the coast of Kuwait. The operation was designed to lay some mines just off the coast of Kuwait the morning of the ground invasion in order to confuse the occupying Iraqis as to the location of the actual invasion. The SEALs had developed a primary egress plan and a backup egress plan for execution after the laying of the mines. If not all the SEALs were back in the special forces boat by a certain time, they were to go to the backup plan in the darkness of the night. Sure enough, when noses were counted on the boat after the mine drop, one SEAL was missing. The team immediately executed the backup plan and soon plucked a surprised and thankful SEAL onto the boat by the back of his collar.

The SEAL team leader's moral of the story: Anybody can come up with a plan; only the good leaders think far enough in advance to come up with a good backup plan designed for contingences when the primary plan falls through. The man with the (backup) plan should be the man in command. Bob had a plan and a backup plan and was in command. Bob did this well. He planned to fly to Orcas Island, unless the weather was worse than expected. If so, he would return. Oh, the 5 Ps? "Proper planning prevents poor performance." Truer words were never spoken about aviation.

Funny weather

"Funny weather we're having today," said the Cowardly Lion. The weather can change very quickly. Bob was wise in getting the current ATIS before taxi. I assume he had already received a weather brief before stepping to the aircraft as well. The tower controller was on the ball in keeping Bob abreast of the soon changing information on ATIS. Depending on the Zulu time reported on the ATIS, this should have tipped Bob off that the weather was changing somewhat rapidly. Remember, ATIS information is usually changed every hour, often just prior to the next hour (around the 55-minutes-past-the-hour mark). If the controller was giving an update to ATIS shortly after the last recording (say, less than 15 minutes), it means that conditions are changing fast. The visibility had dropped from 10 to 7 mi.

For your eyes only

Bob also used an old trick that we too often forget about with all our fancy radio, radar, and other automated information. He did his own PIREP (pilot report). He noted that an aircraft had just shot an approach to Runway 16, circled to land Runway 34, and was never obscured by the clouds. This indicated that it was clear for at least a decent distance above the airport.

Don't forget to use you own eyes! It reminds me of the true story of the airline gate agent who was waiting for an aircraft to arrive. The ticket counter called down and asked her if the plane was down on the ground. The agent immediately began banging on the computer keyboard in search of information on the flight. A much quicker step would have been first to take a peek out of the huge glass window overlooking the runway and parking ramp (the aircraft was entering the ramp at the time). Yet she never bothered to look, choosing rather to answer the question only with automation. Again, don't forget to use your own eyes!

The light show

Both Bob and the controller did a nice job with lighting equipment. When Bob entered IMC unexpectedly, he immediately turned off the landing/taxi lights and strobes. These lights can be extremely disorienting when in IFR conditions. It's like being inside a gigantic ping-pong ball with a disco light and strobe. It can make you crazy. The remedy as demonstrated by Bob: Turn out the lights, the party is over!

The tower controller also used his smarts. When Bob called up to report IMC and a return to Bellingham, the controller immediately and without asking turned up the airport lighting to maximum. More than one pilot has located a field in "heavy soup" by the use of bright lights.

Airport lighting can sometimes cut through the soup. Of course, the key is once the aircraft breaks out and has the field in sight, the controller or pilot (if they can be pilot controlled through the radios) must quickly dim the lights so as not to cause discomfort to the eyes and disorientation to the pilot. This young controller did well on that count. There is one count that he missed, however.

Notice when the controller told Bob about his broken ceilometer—after Bob was back on the ground. It would have been nice to have known about this equipment malfunction before Bob took off. A ceilometer is simply a device that is able to ascertain the bottom of the cloud decks. With a functioning ceilometer, tower can give a pilot a fairly accurate estimate of the actual ceiling so that the pilot can determine if VFR flight is advisable. It is interesting in the exchange about the rapidly changing weather conditions that the ceilometer OTS was not mentioned. Of course, controllers make mistakes, a subject that we will visit in depth later in the book.

Treetop flight

Things got dicey when Bob tried to return to the field. To use his own words, "My instrument scan was poor, and I realize now that I was spending too much time looking outside for ground references. I had leveled the wings but was still descending when I saw treetops by the light of the navigation lights." Remember when I said that we would return to the subject of proficiency? We are there now. Bob said his instrument scan was poor. That can mean slow or incomplete or both. A slow instrument scan means simply being lethargic about covering the outside and inside references in a timely fashion. A timely scan allows you to catch errors or trends quickly before they become gross errors that can lead to trouble. An incomplete scan is a failure to attend to all the relevant references, both inside and outside of the aircraft, including the horizon and clearing for other aircraft outside of the cockpit and the attitude indicator, airspeed indicator, altimeter, and vertical velocity indicator (VVI) inside of the aircraft. While it is not clear if Bob really meant his scan was slow, it seems clear that it was incomplete. Bob was so engaged in looking for the airport that he failed to clear around the aircraft for obstacles (such as trees) and failed to monitor his VVI and altitude as he descended toward those trees he didn't see. The cause of a poor instrument scan is a lack of proficiency. The best remedy for a poor instrument scan is practice, practice, practice. Granted, besides practice a pilot needs to have good scanning patterns ingrained. Gopher et al. (1994) have shown fairly convincingly that pilots can be trained to scan better. But these trained scanning strategies can only be maintained through practice. Practice leads to proficiency. It seems clear that Bob wasn't practicing very hard up until this night. When you do some day VFR flying, bring along an observer and get some "hood time" to keep up the instrument scan proficiency.

Note that Bob had a passenger on board. In this situation, you should use the passenger's eyes. The passenger can help look (read: fixate on looking) for the airport while you scan. Or you can have the passenger back you up (if a qualified pilot) on your scan and clearing while you focus more on locating the airport. This works very well if the airport is on your side of the aircraft.

Channelized attention

This "spending too much time looking for the airport" is known as *channelized attention.* Channelized attention occurs when a pilot focuses on one aspect of the environment to the exclusion of other relevant cues. Channelized attention is also known as *cognitive tunneling* or *fixation.* We all know what it is like to fixate or stare at something intently. Cognitive tunneling is the idea of mentally bearing down on something. We can become obsessed with something and fail to think about anything else. For instance, say you are walking through a store intently staring at something, which causes you to bump into another customer. It is startling when you realize what you have done. Unfortunately for Bob, he wasn't just running into another customer. Trees are much less forgiving.

Reactions

Bob's reaction when he saw the trees was right on target. He did an immediate pull up and power increase. There is no time to hesitate in these situations! You must react immediately, and Bob did—nose up and power up. You must get tracking away from the ground or other obstacle. Time is of the essence. After the pull-up began, he felt the shudder. Without an immediate pull-up, he would have felt more than a shudder.

He suspected that the left gear had impacted the tree. Interestingly, he felt "that the only airworthiness concern would be upon landing." There should have been other concerns, as a host of problems could have manifested themselves. Probably the most severe potential problem would be a large branch or other part of the tree being lodged in the gear. A large branch could cause the small aircraft to tip on its side during landing—not a pleasant thought. Similarly, the tree could have punctured the left-side wheels of the gear so that the tire was flat, or if the tire was just severely bruised, it could blow out upon impact with the concrete. Either of these scenarios could cause a ground loop (in this case an immediate left 180-degree turn because of the drag being located on the left). At higher landing speeds a ground loop would likely cause the aircraft to roll or flip, which is not a pretty sight. Finally, depending on the aircraft braking system, the impact with the tree could have severed a hydraulic brake line to the left landing gear. A severed brake line would obviously cause the left brakes to fail. With only the right brakes operating, the aircraft would experience drag on the right side and not on the left. This would make aircraft directional control on the runway a very tricky proposition for the pilot. A ground loop or departure off the right side of the runway or failure to stop within the length of the runway are all real possibilities. So what should Bob do with these possible scenarios?

An option pilots have taken over the years is to execute a tower flyby so that the tower can inspect the main landing gear for possible malfunctions. A severed brake line may be difficult to detect visually, but sometimes hydraulic or brake fluid can be seen streaming from the gear. Bob could also have indications of a leak aboard the aircraft, depending on instrumentation. A flat tire or a piece of the tree lodged in the gear would be more easily spotted by the tower. Often military bases have a supervisor of flying (SOF) on duty. This experienced pilot is usually stationed in a mobile unit near the runway or in a truck. This pilot is also a candidate to look over the aircraft for damage. A chase aircraft flying in formation or quickly dispatched from the ground is also a possibility for a second opinion on damage to the aircraft. Bob did not have either of these options, but you might. Keep that in your back pocket.

Bob also had two other things going against an inspection by the tower. First of all, it was nighttime—late enough in the evening that it was very dark, depending on moon brightness that night. Some facilities have search lights available. Coupled with lighting on the ramp, there may be enough light to accomplish the inspection. At a smaller facility like Bellingham, this is probably unlikely. Even with enough light, Bob had the quick-moving weather to contend with. Depending on how quickly it was moving in, Bob may not have enough time to accomplish the flyby and still maintain visual contact with the runway in order to land. This would also be influenced by instrument rating and whether the airport had an operable instrument approach available. By choosing to ignore these options, Bob was rolling the dice that the landing would be uneventful. Don't take this chance if you don't have to.

Summary

As most pilots realize, weather is not to be messed with. The forces of nature are very powerful. We have improved our ability to deal with the weather since the days of the Wright brothers, but we are still not to the point of true "all-weather capability." Remember, *always* get a good weather briefing before flying and don't play games to get a good one!

Craig (1992) has done some excellent research on the type of pilot most likely to have a weather-related incident. Many of the factors he lists can be found in the cases we have covered. If you have to turn around in flight because of weather, realize that is part of the flying business. As Chip found out, sometimes we just can't get there. The decision to continue to press or turn around is influenced by knowing the limits of your aircraft, your company's or organization's regulations and SOPs, and federal regulations. Finally, you should know your personal limits and they should be in line with the aforementioned boundaries. Another factor in determining if it is time to turn around is the condition of your instruments and radios. If they are acting up, turn around! The weather isn't going to improve your radios and instruments, though it could certainly make them worse. If you do turn around, don't let the foiled plans become emotional jet lag or baggage. If you want to grieve about it, do it on the ground after the flight. There will be plenty of time then.

The Boeing 727 trying to get to Belize realized that flying in unfamiliar territory offers its own thrills. The destination airport may not have the comforts of home you are accustomed to, especially outside of the United States. If the winds are random and strong at a location, consider holding if time and fuel permit. Often once a storm blows through, the airport will have calm conditions. In the case of thunderstorms in the area, use your weather radar on board (if available) to identify the most menacing cells. Sometimes Flight Service radio or other facilities can provide this information if the capability does not exist on the aircraft itself. Carry an extra amount of airspeed on final in wind shear or gusting conditions. One extra knot for every knot of gust is a good rule. If you do encounter wind shear, take immediate action. Use at least go-around power (or more if needed) and get the nose tracking upward (without stalling the aircraft). Then if conditions at or near the airport make a safe landing questionable, go to a suitable alternate when available and if time and fuel conditions permit.

Finally, we learn from Bob that channelized attention can get you. Keep a good visual scan going at all times. Inside and outside of the cockpit, keep the eyes moving. Have any passengers help either look for the airport (or other target) while you fly. Flying the aircraft is the first priority. Have a backup plan when you take off and as you fly. Plan for contingencies; think them up when you have "drone time." You never know when you may need one. Remember the 5 Ps and practice them! Try your best to stay proficient in your aircraft. If you have lost proficiency, invite an experienced pilot to fly along until you get up to speed. If that is not possible, be very particular about your first flight, making sure the weather is very good.

Always use the resources around you. Rapidly changing ATIS reports indicate fast-moving weather coming in. Observe other aircraft before you take off. Use your available resources well; aircraft lighting and the airport lighting can help or hinder. If you do strike something, try to have the tower or another aircraft look over your aircraft before trying to land (assuming controllability is not a factor). Finally, if you see an obstacle coming up fast, do as Bob did: Pull up immediately, getting tracking away from the obstacle or ground as you add power. A quick response saved Bob.

My ultimate appreciation for the weather came as a 12-year-old, scared, crying my eyes out, and praying like crazy laying on the concrete floor of my basement under a big wooden table. When I later ascended the stairs, I witnessed what can happen to concrete blocks and bricks when weather bears down. I witnessed the complete disappearance of the top half of a soundly built house. Concrete blocks and brick are no match for a tornado. I was amazed—at the power of the wind.

In the next chapter, we will consider a special subset of weather-related problems: VFR flight into IFR conditions.

References and For Further Study

Adams, M. J., Y. J. Tenney, and R. W. Pew. 1995. Situation awareness and the cognitive management of complex systems. *Human Factors*, 37(1):85–104.

Bellenkes, A. H., C. D. Wickens, and A. F. Kramer. 1997. Visual scanning and pilot expertise: The role of attentional flexibility and mental model development. *Aviation, Space, and Environmental Medicine,* 68(7):569–579.

Craig, P. 1992. *Be a Better Pilot: Making the Right Decisions.* Blue Ridge Summit, Pa.: TAB/McGraw-Hill.

Gopher, D., M. Weil, and T. Bareket. 1994. Transfer of skill from a computer game trainer to flight. *Human Factors,* 36(4):387–405.

Kern, T. 1997. *Redefining Airmanship.* New York: McGraw-Hill. Chap. 4.

Reason, J. 1990. *Human Error.* Cambridge, England: Cambridge University Press.

Shappell, S. A. and D. A. Wiegmann. 2000. The human factors analysis and classification system—HFACS. DOT/FAA/AM-00/7.

Wickens, C. D., S. E. Gordon, and Y. Liu. 1998. *An Introduction to Human Factors Engineering,* 3d ed. New York: Addison Wesley Longman, Inc.

3

VFR to IFR

This chapter focuses on a special subset of weather phenomenon. Many aviation experts consider this the most common error among general aviation pilots: flying under visual flight rules into instrument meteorological conditions. It is the second leading cause of accidents among general aviation aircraft, next to loss of basic aircraft control. Way back in 1955, AOPA (Aircraft Owners and Pilots Association) did a study on how effective non-instrument-rated pilots are in handling IMC conditions. Of the 20 pilots studied, 19 placed the aircraft into a "graveyard spiral," while 1 put the aircraft into a whipstall. The "best pilot" lasted 8 min, the worst 20 s (Bryan et al. 1955). VFR flight into IMC is a recipe for disaster. As our Bay Area pilot will soon discover, flight over water can exacerbate this condition.

Flying by the Dock of the Bay

"It was football season in the Bay Area. Steve Young and the 49ers were playing in Candlestick that weekend. The

Silver and Black of Oakland were preparing across the Bay. The air was crisp, as it often is on a fall Friday morning in the Bay Area. Commuters were beginning to make their way into the city. I was preparing to take off early that morning. Dawn was breaking, it was just after six, the sun was up, and so was I. I tuned up the San Carlos ATIS (Air Terminal Information System). It reported scattered at 800 ft, haze, broken at 10,000 ft. I rolled my small Cessna down the taxiway. I took off and crossed the bay coastline on a east–northeasterly 060-degree heading. I stayed below the clouds at 800 ft and turned toward the San Mateo bridge because it looked clear in that direction. At about 2 mi from the bridge, the haze became clouds (or was it the Bay Area fog?). Suddenly I was in IMC. I immediately started a descent and began a 180-degree turnback to the airport. At 400 ft, I had not broken out, so I leveled off. I counted to 10, then began a 500-fpm climb (I am not instrument rated!). I broke out on top at 1200 ft. I didn't know my position. I flew 060-degree, 090-degree, and 180-degree headings. Finally, part of my senses returned. I tuned up the Global Positioning System (GPS) on Livermore Muni (LVK). When it locked on, I flew toward Livermore until the undercast was gone. I had flown to within 5 mi of San Jose International (SJC). I could see planes coming up through the undercast before I turned toward LVK." (ASRS 417742)

Post-game wrap-up

I think we can all agree that this pilot was lucky, if you believe in such a thing as luck. Fortunately, he didn't enter the San Jose traffic pattern area where a midair collision would have been more probable. Fortunately, he didn't get into some sort of spatial disorientation such as the coriolis effect or a graveyard spin. Fortunately, in his attempt to escape from IMC, he didn't impact the water in controlled flight. Why was he so fortunate?

First of all, he did what he could to analyze the weather, assuming he got a good weather report before stepping into his aircraft. The report is not clear on this fact. However, the pilot did dial up and receive the current ATIS information. Assuming he also did a visual scan of the area from the ramp and during taxi to takeoff, he was probably pretty safe in taking off. Here is where the problem begins.

After takeoff he remained clear of the clouds and headed for what seemed to be a clear area (toward the San Mateo Bridge). That is where it got tricky in a hurry: "The haze became clouds." Suddenly, he was in IMC, and he was not IFR-rated. It can happen, weather is funny in that way; it's not very predictable. Witness the fact that weather forecasters on TV are often wrong, yet they remain employed! That's because weather can change fast. At this point the pilot puts himself in a pinch by starting a descending 180-degree turn. I'll give him the benefit of the doubt and assume he thought he would quickly pop out under the haze/clouds, but he had two better options.

Spatial disorientation

The first option was to begin a *level* 180-degree turn-back toward VMC. The area he was just in was VMC, so he could've returned there. The advantage of the level turn is that it reduces the possibility of spatial disorientation. By executing a descending turn, he was stimulating the semicircular canals in the inner ear in more than one plane, increasing the chances of spatial disorientation. Situational awareness is closely related to spatial disorientation and is the subject of another book in this series, so I don't want to go too far down this rabbit trail. But allow me to give a quick primer on how a pilot senses motion, which will be review for many of you. There are three main components: the semicircular canals, the otolith organs, and the somatosensory system. As you remember from high school biology (you do remember, don't you?) and from flight training, the *semicircular canals* allow the human body to detect angular motion in three dimensions. These canals are intricately designed at right angles so that one is aligned with the horizontal, vertical, and lateral (x, y, and z) planes. The fluid inside the canals stimulates special hair filaments or hairs, which, in turn, send messages to the brain indicating angular motion. The semicircular canals can report rotation in three dimensions. That's the simplified version. In close proximity to the semicircular canals are small sacs known as the *otolith organs*, which detect linear and radial acceleration through the bending of small hair filaments. Finally, the human body is equipped with the *somatosensory system*, which are sensors embedded in our skin, joints, and muscles. These sensors are responsive to pressure and stretch, and they help us to maintain our equilibrium. Basically, these sensors are what give the pilot the seat-of-the-pants sense. We are taught in flying not to trust these sensations, but rather to rely on our visual system and our instruments—especially in IMC conditions.

The danger in executing the descending 180-degree turn is that the semicircular canals would be stimulated in more than one axis (coupled with a sensation of radial acceleration in the otoliths), which can induce spatial disorientation and result in one of many harmful illusions known as *somatogyral illusions*. One such example is the *coriolis illusion*. Coriolis is considered by many to be the most dangerous of all illusions because its overwhelming disorientation power. This effect can occur during a constant descending turn or holding pattern if the pilot makes a quick or jerky head movement in any plane different from that of the aircraft. This can cause the sensation of movement in a plane of rotation where no real motion exists. Often the pilot will feel the overwhelming sensation of a roll with a climb or dive. This is the coriolis. Often the pilot will make an improper control input to adjust for a condition that does not exist, which will result in impact with the ground. It is imperative to avoid quick head movement while turning (and especially descending) in IMC. As should be apparent from this description, our Bay Area pilot was set up perfectly for this illusion. Had that pilot jerked his head, he likely would have been a part of the San Francisco Bay. Again, that is not the subject of this book, but beware that such possibilities exist and the outcome isn't pretty.

There was also a second option, which was a climbing 180-degree turn. The downside of this approach is similar to the descending turn: the semicircular canals would be stimulated along with the otoliths, with a real possibility of spatial disorientation occurring. Again, the spatial disorientation could result in an illusion and erratic aircraft control. However, this option at least offers a big upside. The aircraft is now pointing away from the ground and in this case the ice-cold San Francisco Bay. Altitude is usually the pilot's friend, and in this situation it's a good friend.

Wind the clock

At this point we need to give this pilot a kudo. At 400 ft AGL, he realized he was in a bad position, and it was getting worse in a hurry. Remember, he said, "At 400 ft, I had not broken out, so I leveled off. I counted to 10, then began a 500-fpm climb (I am not instrument rated!)." He stopped the progression. Later in this book I will discuss the concept of the error chain. As many studies have shown and many a pilot recognizes, usually it's not one

error that gets us. It is a series of errors. This pilot was in
the early stages of such a series of errors. He had entered
IMC, and knowing he was not instrumented-rated, he
elected to execute a descending turn. At this point he was
heading toward the water. Then a red flag appeared mentally. He stopped everything and got back in control.
Many pilots have been taught the three basic rules of
handling an emergency: (1) Maintain aircraft control;
(2) analyze the situation and take proper action; (3) land
as soon as conditions permit. Step 1 is the key: Maintain
aircraft control. That is what this pilot did. He stopped the
descent and leveled off. Then he counted to 10; some call
that "winding the clock"—settle down, wind the clock,
and think. That allows you to analyze the situation and
take proper action. He analyzed the situation and decided
that a controlled 500-fpm climb was best. After 2½ min,
he had broken out on top of the clouds. Now he was back
in business—the business of figuring out where he was.
Notice he said, "Finally, part of my senses returned. I
tuned up GPS on LVK." I have been there before, and
I am sure you have too, where you just aren't focused,
and then the lightbulb clicks on. It is one of those
"eureka" experiences: "Hey, I have a GPS. Let's use it."
Once he was locked on to the Livermore (LVK) location,
he could move away from the San Jose International
Airport. As the old expression goes, "Aviate, navigate, and
then communicate." The pilot demonstrated these first
two steps well.

Anchor and adjust

Why is it so common for non-instrument-rated general
aviation pilots to enter IMC? One reason is a phenomenon known as "anchoring and adjustment." It is a well-
researched decision bias that we have as humans. Let me
illustrate. If you ask some average people on the street,

"How long is the Mississippi River?", they will give you some figure, let's say 1400 mi. But if you ask some other people, "Is the Mississippi River greater or less than 500 mi long?", most all of them will realize that 500 mi is clearly too short a distance and answer, "Longer than 500 mi." Yet if you ask them to then estimate the actual distance, they will answer with a figure much lower than that of the first group. They will average around 900 mi in their estimate. The figure that you gave the second group (500 mi) acted as an *anchor* and further estimates (900 mi) were *adjusted* from that anchor. Meanwhile, the first group had no such anchor to begin with; therefore, it had a much more accurate estimate of the actual river length (which is 2350 mi, by the way).

You may never have had the experience of someone asking you the length of the Mississippi River, but we have all encountered this phenomenon when making purchases, especially large purchases. The car salesperson is the classic example. He or she puts a retail sticker price on the car and then "knocks it on down" from there so that you "know you are getting a great deal." Of course, the original sticker price was absurdly high, but it effectively serves as an anchor for those adjustments, which make us feel so much better. Now that accurate wholesale prices are readily available, this technique is not nearly as effective. This technique is also often used for that odd product advertised on late-night TV: "How much would you be willing to pay for this valuable product? $100? $200? Actually, this product is yours for the incredibly low price of $19.95. But that's not all—if you order now..." The manufacturer knows the product isn't worth $200 or even $100, and so do you, but it sets a high anchor, and then adjustment takes place.

Nice business lesson, but what does this have to do with flying? Specifically, what does it have to do with

VFR-to-IFR flight? Actually, quite a lot. The original weather forecast we receive acts as an anchor. In this case, from the ATIS information and from a visual scan of the immediate area, an anchor of "good VFR weather" was set. The pilot made adjustments from this anchor. Let's suppose the ATIS would have advised heavy fog in the area. The pilot would then adjust from that much gloomier forecast, perhaps elect to not take off, or, if he did take off, turn around and land on the first indication of any fog at all. So our initial weather brief has a big impact on how we look at weather once airborne. Perhaps it is good that many weather forecasters tend to be a bit conservative when making weather calls. Some people say that this is just to cover their rear ends in case of trouble. There may be some truth in that statement, but it has an added benefit for the pilot. It allows us to establish a more conservative anchor to begin with. Of course, if the pilot "games the system" by saying, "Ah, he's always forecasting the worst," then the pilot has simply moved the anchor toward the more risky side. Furthermore, if the pilot receives no weather brief and simply tunes in ATIS, he or she will get actual conditions at the time of recording, not the forecasted conditions. Therefore, the pilot may be set up if bad weather or IMC is moving in and a poor anchor has been set. So beware and base your decisions on what you see, not on what the weatherperson told you.

Dig me a tunnel

Sometimes a pilot's attention can get too focused. This condition is known as *channelized attention, cognitive tunneling*, or *fixation*, as discussed in the previous chapter. It usually involves focusing on a portion of the cockpit or outside cue to the neglect of other important information. The Bay Area pilot seemed to "lock in" as he "turned toward the San Mateo bridge because it looked clear in that direction." My question is, did it look clear anywhere else or what did the area around the bridge look like? In fairness to the pilot, he can't address the question and he may well have scanned the entire area. However, for the sake of the argument, let's assume he looked only at the bridge. When we focus in on a certain cue and ignore conditions surrounding that one piece of evidence, we can get hemmed in quickly. Cognitive tunneling can cause you to miss the big picture, and the big picture may hint at trouble if we take the time to examine it.

Scud Runner

"While en route from Texarkana, Arkansas, to Coleman, Texas (COM), we were on an IFR flight plan. I was flying my C310 with a passenger aboard. We encountered solid IMC around Cedar Creek VOR, but as we passed Glen Rose VOR, the weather at our cruising altitude of 6000 ft began to clear. I had been briefed that Coleman NDB (nondirectional beacon) was OTS (out of service) and that forecast conditions for our arrival would be a ceiling of between 1500 ft and 2000 ft. Between Cedar Creek Vortac (CQY) and Glen Rose Vortac (JEN), I radioed flight watch, and it indicated that at Abilene (ABI) and San Angelo (SJT), the closest reporting points, VFR conditions prevailed. A commuter flight departing Brownwood indicated that there were a few clouds at 1500 ft and that the ceiling was about 3000 ft. Knowing that the MVA (minimum vector altitude) in the area was 4000 ft or 5000 ft and knowing that I would be unable to do the NDB approach at Coleman, I knew that our only chance at getting into Coleman would be to proceed VFR under the overcast. Just northeast of Brownwood, there was a large hole in the clouds, through which I could see the ground. I decided to cancel IFR and descended VFR. Center told me to maintain VFR. I remained with Center for advisories because I was still 25 mi from my destination. I descended and got below the clouds at 2000 ft MSL (600 to 700 ft AGL). I maintained VFR in class G airspace. The weather had begun to deteriorate and conditions were worse than both the forecast and the PIREP (pilot report) I had received.

"As I headed over Brownwood Lake, I realized that I would be unable to maintain VFR for much longer. I was unwilling to descend further; I radioed Center and requested another IFR clearance and a climb. I was told to stand by. At that time, I entered IMC. Since I was so close to the ground and conditions had changed so rapidly, as I waited for my clearance I decided to start a climb away from obstructions. At this point, I figured that climbing straight ahead into IMC was safer than attempting a low-altitude 180-degree turnback to conditions I was unsure were still VMC. As I passed 2500 ft MSL, Center cleared me back on course with a climb to 5000 ft. I ended my flight with a localizer approach into Brownwood Regional Airport (BWD).

"There were, to my knowledge, no further problems. To the best of my knowledge, I did not enter controlled airspace before being cleared to do so. While it would have been best not to have ever gotten myself into the situation in the first place, once presented with the situation, it was safer to climb rather than attempt a low-altitude 180-degree IMC turn over unfamiliar terrain. I was in contact with Center the whole time and was certain that there were no other aircraft in the area and that I presented no hazard to other aircraft as I climbed. ATC (air traffic control) was not put out by my actions and there were no further incidents during the flight. In the future, I realize that it will be better to land at an alternate airport, serviced by an operational approach, than to attempt to fly under clouds, in marginal conditions, to my intended destination." (ASRS 423268)

Contrast

Let's contrast our Texas pilot to our friend back in the Bay Area. Both pilots unexpectedly entered IFR conditions, but there were a few key differences. The Texas pilot seems to have been instrument-rated, since he filed an IFR flight plan and mentioned shooting a localizer approach and did not mention the fear of being VFR-only rated when entering IMC. This difference between the two pilots is not to be minimized. Not surprisingly, it seems that the majority of general aviation pilots entering IMC from VMC that end up as CFIT or CFTT cases are not instrument-rated pilots, so upon entry to IMC, panic can set in. Panic is not the pilot's friend. Though the entry may be unexpected, at least the instrument-rated pilot has been trained to deal with it, once the pilot has regained his or her composure.

Another key difference is how each pilot chose to control his aircraft upon entry into IMC. The Texas pilot chose a straight-ahead climb away from uncertain conditions and away from the ground. Contrast this with the Bay Area pilot who chose a descending turn. One reason for this difference is the instrument rating. Being instrument-rated gives the pilot the confidence to climb straight ahead into the IMC, rather than risking a descending turnback toward possible VMC. Instrument rating makes this an option. Remember the lesson from our Bay Area friend. It is always best to climb away from the ground. The Bay Area pilot may have been wise to choose the straight-ahead (versus a turn) climb as well to avoid possible spatial disorientation. However, climbing away from the ground, as illustrated by the Texas pilot, is key.

Get-home-itis

Most every pilot who has flown for any amount of time is familiar with the phenomenon of "get-home-itis." Get-home-itis is also known as "get-there-itis" or simply "pressing." It is when your desire to arrive at a certain destination (often home) overrides sound judgment, and inappropriate risks are taken. We have all experienced this while driving a car. Perhaps you are late getting home or picking up the kids. You press, and perhaps that bright orange light you just sailed through was just a bit more red than you are willing to admit. The Texas pilot does not directly attribute his circumstances to get-home-itis, but it is inferred in his analysis of the situation that he should have gone to a suitable alternate with a serviceable approach rather than pressing in marginal VFR conditions trying to make the intended destination. He wanted to get to his destination, whether it was home or otherwise. The remedy for get-home-itis is actually rather simple. Be honest with yourself. Must you really get there at the risk of your life and your passengers' lives? The answer is almost always no. In economics, we call this a cost/benefit analysis. Do the potential benefits of this operation exceed the potential costs of this operation? If the answer is no, then you shouldn't do it. This cost/benefit analysis is actually the remedy to many pilot judgment errors. But at times it takes extreme self- or flight discipline to resist the urge to press on to the destination. One of the few exceptions to the risk issue is a military mission of extreme national security or tactical importance. Then the urge to press need not be tempered so greatly.

Perhaps the ultimate example of get-home-itis that I have heard or read about (and there have been lots of these stories; it seems we all have one) comes from a former Air Force wing commander at Reese AFB in Lubbock, Texas. There were a number of pilots in the room for an annual instrument refresher course, when the topic of pressing came up. The full colonel relayed a story from his young days as a captain in the 1960s. It seems that two of his squadron mates, both young captains, were scheduled for a cross-country sortie. This is a sortie where two instructors are allotted cross-country time to hone their proficiency, as instructor pilots can get rusty in actual flying of instrument approaches after spending much of their time verbally instructing the students. The added benefit of these flights (other than the proficiency) was the opportunity to RON (remain overnight) at another base. These two young captains targeted Randolph AFB in Texas, as there were two young ladies there they were eager to rendezvous with. As is often the case, when you *really* want to go somewhere, that is when you have car trouble, and airplanes are no exception. They had to wait until late Friday afternoon, until the student sorties were finished, to depart from their home base. When they got in the T-37, they could only start one engine. The other jet engine would not crank. After consulting with the crew chief, they gave it another try, to no avail. They desperately wanted to take off, but they needed both engines to be running to do so. This was by regulation and by common sense.

After conferring for a while and as the sun set deeper and the Randolph ladies seemed more and more distant, they came up with a great idea—or so they thought. They decided to take off with just one engine operating and then accomplish an air start on the other engine, which they were sure would work. These two Romeos actually executed their plan—with just one glitch. Once airborne (and they did get airborne), the lone good engine flamed out and they crashed. This is one of the all-time-great acts of air stupidity, all because of get-home-itis.

Training Flight

The supersonic T-38 Talon took off from the training base in the southwestern United States. Aboard was a young first lieutenant instructor pilot sitting in the back seat. An eager second lieutenant occupied the front seat of the sleek training jet. The T-38 was the second aircraft flown in the 11-month undergraduate pilot training (UPT) program. Today's mission was a VFR low-level near a range of mountains.

As the aircraft approached the route, the IP noticed that there was some cloud cover that day on the route. This was rare, as Arizona has so many days of sunshine that the IP rarely had to be concerned with poor weather. The IP was unsure whether they would be able to maintain VFR flight but directed the student to descend down to under 500 ft AGL to enter the route. The aircraft canceled IFR clearance with Center.

As they flew they encountered increasingly cloudy conditions, and the IP ordered the student to pop up out of the route. The aircraft contacted Center to regain its IFR clearance. Suddenly, the aircrew canceled the request, advising ATC that it was reentering the route. Center gave the approval, and the aircraft descended back into the route. Several minutes later the aircraft impacted a mountain, killing all aboard.

Young and invincible

The facts of this case are fairly straightforward. What is difficult is our inability to read the IP's mind. There were no mechanical problems with the jet. The accident investigation board ruled it as a case of pilot error and categorized it as a CFIT. But why would the crew attempt to fly the low-level route in questionable conditions? We can never know for sure, but we can speculate from variables that we know to be true. I emphasize that this is speculation. My intent is not to demean either of these crew members.

One likely scenario is that the IP may have suffered a little bit from the "Superman syndrome." That is where the pilot feels that he or she is invincible. Accidents only happen to the other guy. This malady afflicts mainly the young; the old have had enough close calls to be a bit wiser. This IP was about 25 years old. He had done well during UPT and had been selected to return as a first assignment instructor pilot (FAIP). Many FAIPs had hoped to get fighters right out of UPT; they still had a chance for a fighter after their instructor assignment was up. But if you can't fly a fighter, the T-38 is the next best thing.

This FAIP had done well as a student and now was flying the T-38, so he had reason to be proud. However, his youth may have tempted him to take a risk on the low-level route because he may have been a bit too proud, a little too invincible.

Organizational factors

Military instructor pilots are not lone rangers. They work for and are influenced by organizations—in this case a flying training squadron. Safety is the primary concern of all undergraduate pilot bases. A true concern for these young men and women exists. However, safety can have a way of fading into the background. The second-highest priority in undergraduate pilot training is the time line. A new class starts every three weeks. It is imperative to keep that class on pace to graduate in the 11 months allotted to it.

Safety is briefed everyday to the higher echelons. If there are no accidents or incidents, then the briefing moves on. It moves on to the time line. Each class's time line is briefed and highlighted. Each flight commander knows that a good portion of the judgment of how well he or she is doing on the job is reflected by that time line. I clearly recall my flight commander after the second week of flying training giving a "high five" to the assistant flight commander: "We are already three days ahead of the time line after two weeks!" It was the happiest I saw him during all of pilot training. Could the time line have been going through the IP's mind? Perhaps he was thinking, *I've got to get this guy his low-level today. He's getting behind. We need this sortie.* We can't know for sure, but I am confident that the IP was aware of the time line, whatever its state was, before he took off. It had been briefed.

The availability heuristic

Our decisions are often influenced by what we can remember. What we can remember the easiest is known as the *availability heuristic.* Yet what we can remember most easily may not be the best response or the most likely explanation; it just comes to mind easily. If this explanation has proven successful in the past, it is even more likely to be recalled. For example, let's say the IP saw the clouds and remembered that during his training he once successfully flew a low-level through similar conditions. That fact would make his decision to go again that much easier. Or he may have heard Joe talking over lunch the other day about how he did a similar low-level. Or perhaps he would have remembered the time line if it was emphasized that morning. Conversely, had he just been briefed on a low-level crash, the crash would likely have come to mind because it was the most available. In such a case, the chances are the IP would not have risked the low-level.

The remedy for the availability heuristic is to not stop thinking after one option has come to mind. Force yourself to consider multiple options. This may require effort on your part, as important information may not be readily available to your memory.

Summary

As we have seen throughout this chapter, VFR flight into IMC can be dangerous for the civilian as well as the military pilot. There are lots of factors that can lead to this condition, such as leaving the base under iffy weather conditions if you are a VFR-only pilot. This can also be a problem for IFR pilots if they push their lower weather requirements to the iffy region. Once an unexpected encounter with IMC occurs, don't exacerbate the situation by accomplishing a descending turn (or any turn, if possible), which can induce spatial disorientation. If you must turn, then try to do a level turn. Definitely seek VMC as soon as possible, but don't be foolish in your pursuit. A straight-ahead climb may be preferable, especially for your inner ear and gray matter.

Realize that certain cognitive or mental conditions can make you more susceptible. Anchoring and adjustment can set you up to be nonchalant, or overly wary (the riskier of the two being a lack of concern). Cognitive tunneling can lead you to focus on that "clear area" or that "little gap over there." Don't get hemmed in! Of course, beware of our familiar threat, get-home-itis. Remember, count the beans, do a cost/benefit analysis. Is the pressing worth it? Be disciplined. Be a poster child for the Marines, with the mental toughness to say no to the get-home urge.

The organization may put pressure on you to press on with the flight. Keep up with the time line—but remember, safety is paramount! When you encounter IMC, try to remember those who haven't made it. You may have a tendency to remember only the success stories as per the availability bias. Whatever first comes to mind may not be the best route to take.

Speaking of biases, no insult intended, but you are not superhuman; you are not invincible. Remember: Even Superman had his kryptonite. There is a plaque on my wall that reads, "The superior pilot uses his superior knowledge to avoid situations requiring the use of his superior skills." Scud running may be exciting, and it may induce a thrill, but thrills have their downside. Scud running can kill you.

Since VFR to IMC is a problem that particularly plagues the general aviation pilot, I share with you a composite picture painted by the NTSB of the pilot who is most susceptible to this type of accident (Craig 1992). In a nutshell, he or she is an inexperienced, nonproficient pilot, with a taste of instrument flying on a pleasure flight during the daytime with a passenger. I leave you with this question: Is this pilot you? To help you answer, see the entire composite that follows:

1. The pilot on the day of the accident received an adequate preflight weather briefing, most likely over the telephone. The information was "substantially correct" nearly three-quarters of the time. The weather conditions encountered were "considerably worse" than forecast only about 5 percent of the time.

2. The pilot was making a pleasure flight.

3. The pilot had a private pilot certificate but was not instrument-rated.

4. The pilot was relatively inexperienced and typically had between 100 and 299 h of flight time.

5. The pilot was typically not proficient, with less than 50 h flown in the 90 days prior to the accident.

6. The accident occurred during daylight hours, with the pilot flying from visual into instrument conditions either intentionally or inadvertently.

7. The pilot had some instructional instrument time, typically between 1 and 19 h under the hood with an instructor.

8. The pilot had logged no actual IFR time (at least until the fatal encounter).

9. The pilot did not file a flight plan.

10. The pilot was accompanied by at least one passenger.

References and For Further Study

Bryan, L. A., J. W. Stonecipher, and K. Aron. 1955. *AOPA's 180 Degree Rating*. Washington, D.C.: AOPA Foundation, Inc., and Aircraft Owners and Pilots Association.

Craig, P. 1992. *Be a Better Pilot: Making the Right Decisions*. Blue Ridge Summit, Pa.: TAB/McGraw-Hill.

Kahneman, D., P. Slovic, and A. Tversky. 1982. *Judgment under Uncertainty: Heuristics and Biases*. New York: Cambridge University Press.

4

Situational Awareness and Complacency

"Get your SA out of your pocket!" This is a familiar expression in pilot training. It simple means, get your head up, your mind active, and figure out what is going on around you because you are missing it! Kent Magnuson of the USAF Safety Life Sciences Division studied U.S. Air Force mishaps from 1989 through 1995. He found that lost situational awareness because of channelized attention was the single largest contributing factor cited in these mishap reports. It was actually tied with decision making (a closely related topic) as the largest contributor to accidents in the Air Force.

There are a number of formal definitions of situational awareness. Dr. Mica Endsley is perhaps the world's foremost researcher in the area. She defines situational awareness as "the perception of the elements in the environment within a volume of time and space, the comprehension of their meaning, and the projection of their status in the near future" (Endsley 1989). *Human Factors* journal, which covers many topics relevant to safe and efficient flight, defines situational awareness as "the pilot's ability to stay aware of evolving aspects of the flight environment that might become relevant should unexpected circumstances develop" (*Human Factors* 1995). The term is best captured by Crew Training International, which trains many military units in the concepts of crew resource management (CRM). It is clear that good situational awareness is central to CRM, which has the goal of the crew's efficiently using all resources in and outside of the aircraft. Crew Training International defines situational awareness (SA) as the central answer to these three questions:

What has happened?

What is happening?

What might happen?

If the pilot has a good grasp on the answers to these three questions, he or she is likely to have "good" situational awareness. Crew Training International further emphasizes that good situational awareness requires that the pilot be aware of the three physical dimensions (the x, y, and z planes), as well as the fourth dimension of time and how these four dimensions converge. Knowing what is occurring, what has occurred, and what likely will occur in these four dimensions is the essence of situational awareness.

Returning to Endsley's work for a moment, she points out that there are three levels or processes within situational awareness. They have some overlap with the three questions from Crew Training International. The first is the ability of the pilot to *perceive* key features and events in the dynamic environment (e.g., there is motion on the right side of the windscreen I didn't see before; there is a hydraulic gauge that is higher than the others). The second is the ability to *understand* the meaning of those events (e.g., a small aircraft tracking westbound; the hydraulic level has fallen by 1 quart in the last hour). The final process is the pilot's ability to *project* or *predict* the future implications of these changing features (we are on a collision course and I need to take evasive action; at this rate I will lose left hydraulics and the braking system in an hour). These processes represent different levels of situational awareness. To get to the deeper levels, one must have covered the entry levels. What this work tells us is that situational awareness is not an either/or proposition. It isn't the case that the pilot "has" situational awareness or he or she doesn't "have" situational awareness. It is a continuum, and the pilot can fall anywhere along this continuum. The pilot who perceives the change in the environment but stops there has little situational awareness. The pilot who sees the change, understands the change, and is able to project what this means for his or her aircraft has high situational awareness. I emphasize that situational awareness is a cumulative process. To get to the deeper levels of situational awareness, the pilot must have the entry levels mastered. Let's look at some pilots whose lost situational awareness almost cost them their lives.

Medium to Large and a Little Low

"We are categorized as a medium to large transport. I was flying as pilot in command (PIC). We had done a good briefing of the approach en route and put the approach plates away. It was just the beginning of the bad weather season in Texas. We were on approach into Dallas/Ft. Worth (DFW) airfield and cleared for ILS Runway 18R at 3000 ft MSL. We had a load of passengers heading into the Metroplex. Dallas was to our left and Ft. Worth to our right. It was the end of a long day and the fourth day of our trip. We were established on the localizer inside of 10 mi, cleared to descend to 2400 ft MSL. The controller asked us to "keep up the speed" until the FAF (final approach fix) for spacing, so we maintained 180 knots.

We were below glideslope, and I initiated a gradual descent. At 1900 ft MSL it hit me. I realized we were inside the FAF below glideslope and descending! Upon recognition, I immediately initiated a correction back to glideslope, at which time Approach Control made a radio call advising of a low-altitude warning. Glideslope was captured, and an uneventful safe landing was made on Runway 18R." (ASRS 420146)

Why so low?

The PIC who submitted the report did a nice job in analyzing his own problem. I'll cover the good point first. The crew accomplished a good approach briefing during its "drone time" at altitude as it neared the point where the en route decent was accomplished. That takes care of the good point. Oh, I guess we should also commend the pilot for catching the error in flight and mitigating the error with corrective action.

The primary problem with this crew was complacency. Complacency is defined as "a feeling of contentment or satisfaction, self-satisfaction, or smugness" (*The American Heritage Dictionary*). It is this smugness or self-satisfaction that leads us to take things for granted and let our guard down. Complacency is no friend of situational awareness. In fact, they are arch enemies. Unfortunately, two things that often lead to complacency are characteristics of airline crews in particular. High experience and routine operations are precursors to complacency. The more experience we gain, the more confident we feel, which in turn can lead to complacency—"been there, done that." As the old saying goes, familiarity breeds contempt, and it also breeds complacency—that sluggish inattention that we have all experienced both in and out of the cockpit. Complacency is closely related to experience and comfort. Chuck Yeager, "Mr. Right Stuff," says it is the number one enemy of experienced pilots. That is how two instructor pilots flying the two-seat T-37 (note: no student is aboard) are able to land gear up, with a fully functioning gear system.

Complacency can be seen in several places in this study. It was the beginning of bad weather season, so the crew was not spooled up for the approach; they had been doing a lot of visual approaches. Visual approaches are convenient and expeditious for both aircrews and the flying public, not to mention the controllers whose job is to expeditiously and safely direct aircraft to a landing. However, visual approaches can make a crew sloppy. When actual instrument approaches are called for, the pilot can be rusty or not proficient. It can take a while to get back in the swing of things. Furthermore, by the PIC's own admission, they were not anticipating low ceilings, so they didn't even have the approach plate out. This corresponds with Endsley's entry level of situational awareness: perceiving that features are changing in the dynamic environment. To get to the deeper levels of situational awareness, you must have the entry levels. By not perceiving the lower ceiling, they did not understand that an instrument approach was required, and they therefore did not predict the need for their approach plates.

Complacency can often be compounded by physical condition. There is an old sports cliché that says "fatigue makes cowards of all of us." There is a corollary in the aviation world: Fatigue makes slugs of all of us. When we are fatigued we are much more susceptible to complacency or putting our mind on autopilot while our body "does the couch potato." That was certainly the case here, as the crew had been on the road with long work hours for four days. Even when we stay within the company or other organization's crew rest policy, we can be fatigued. Most people don't sleep better on the road, and they don't eat as well either. Couple that with changes in time zone, and your body is much more susceptible to fatigue.

A tear in the space/time continuum

There is a phenomenon known as *time distortion*. Time distortion can occur in one of two ways. Time can become elongated. In everyday life, you have no doubt experienced an occurrence when time seemed to stand still. Something that happened over just a few minutes or seconds seemed to last much longer. Maybe it was during a marriage proposal or that shot in the big game or when you first see that good friend you haven't seen in years—those magic moments. As most know, a similar phenomenon can occur in the cockpit. I've heard pilots describe a hard landing when they really "pranged it on" for several minutes, when the landing probably occurred over a 10- or 15-s period. Yet, they can see every excruciating detail. The same goes for an in-flight emergency, perhaps a near midair collision. Time can seem to be longer.

Time can also seem to become shorter. Known as *time compression*, this is when everything seems like it is in fast-forward on a tape player. You are going like an SR-71 (Mach 3+) mentally. It often seems that we get in this predicament with the help of our good friends, air traffic control (ATC). ATC has a tough job, which is getting tougher; it is called upon to move traffic safely and expeditiously between sectors and to sort out and sequence a growing number of aircraft in and around major airfields such as DFW. The controller here was undoubtedly trying to get the sequencing right at DFW to get as many planes as possible on the tarmac without causing any undue delays for following aircraft. I salute ATC for its efforts; it has a tough job. The result here was a crew rushing the approach and time compression. The crew was well above approach speed (180 knots) as it neared the FAF. ATC also gave the aircraft an abnormal clearance. The crew was cleared to a lower altitude at a critical point instead of just intercepting glideslope at a stable altitude of 3000 ft MSL. Any time we are rushed or given an out-of-the-ordinary clearance, our usual safety habit patterns can be broken. When both occur at the same time, we are particularly susceptible to dropping something out of the crosscheck.

Déjà Vu All Over Again

It was a Tuesday evening. It had been a good flight, and the crew was almost in Denver. Weather conditions there were clear and a million, though it was after nightfall. Frank was the first officer (FO) of the Boeing 727 passenger airliner with about 5000 h, 500 of those in the "27." Ed was the captain with even more flight hours and experience in the 727. They were cleared for a visual approach to Runway 35L at Denver. They made a modified right base and were approaching final at about a 50-degree angle. The captain was flying, descending at a normal airspeed and rate. Everything seemed fine. Both the captain and the first officer fixated on Runway 35R, thinking it was Runway 35L. They had a glideslope indication on their attitude directional indicators (ADIs) with both ILSs tuned correctly to Runway 35L. As they approached the final for Runway 35R, Frank and Ed then noticed at the same time that they had no localizer intercept, and they realized what had happened. They both looked at Runway 35L, and Ed started correcting to intercept the final for Runway 35L. Just then, the tower controller said "Low altitude alert; climb immediately to 7000 ft!" They were at about 6000 ft at the time. (ASRS 420927)

Can you say the "C" word?

Again in this scenario we see the "C" word (compla-cency) associated with a lack of situational awareness. Frank, the first officer, made the report. He puts it better than I can: "Causes: poor situational awareness, com-placency, a very poor showing with no excuses at all. Could have been fatal at another airport. P.S. We never got close enough to get a GPWS warning, but I'm sure we were in the vicinity."

Actually, the complacency likely caused the poor situ-ational awareness. If this Boeing 727 was operated by United Airlines (there is no way to tell for certain), it is likely that this aircrew had flown into Denver literally hundreds of times, as Denver is a major western hub for United. In fact, it houses United's training center so crews are in and out of there a lot, even when not flying pas-sengers. As I said earlier, familiarity breeds complacency. Flying into Denver could have been old hat, and with nice weather, the mind was not geared up for action. Both the captain and first officer missed it. Who knows where the second officer (the flight engineer) was men-tally. He clearly wasn't monitoring the approach very well. Six pairs of eyes were looking at the wrong runway.

However, these crew members practiced a good habit that saved their rear ends. Dialing up the localizer on a visual approach is always a good backup. In this case they saw no localizer intercept close in to the runway when it seemed they were lined up. It triggered the "on switch" in their minds. The wrong runway! Fortunately, there was time for a correction. Fortunately, there was no parallel traffic to Runway 35R, or a midair conflict could have been imminent. The controller was helpful with the low altitude alert. The climb, missed approach, and new approach were the best way to go. Start fresh and try it all over again.

This case could launch a discussion of CRM, but I will save that for another chapter. Similarly, the topic of automation and GPWS plays a part, but this too will be covered later in the book. What is important to notice here is that Frank and Ed's Boeing 727 had a loss of situational awareness, which resulted in a deviation right of course (and then an altitude problem). Either way you slice it, lost situational awareness can affect you in the three physical dimensions, as well as the time dimension.

Do the Hula and Slide Over

I want to be careful not to cast stones at Frank and Ed. I had a similar experience to theirs flying into Honolulu International Airport in the mighty KC-135. Hickam AFB is collocated with the Honolulu Airport, and we were fortunate to be heading there on a mission. As many of you know, the city of Honolulu is only 3 mi east of the International Airport and is sensitive about noise. They take noise abatement procedures very seriously. It is a way of life. They don't want the tourists scared away. As a result, takeoffs at that airport are normally made from the Reef Runway (26L/08R), a runway built in the late 1960s on an artificial reef out in the harbor. Immediately after takeoff, aircraft are expected to begin a shallow turn out to sea and away from tourists. Getting into the airport is another matter. Normally, the approaches curve you into toward two parallel runways on land (04R&L/22R&L) from out over the ocean. You have little time to line up with the runway under VMC. As a result, you need to have a good visual on your runway as soon as possible in order to line up properly. There is likely to be several jets stacked up behind you, so the tower expects no delay.

As a crew (aircraft commander, or AC, copilot, navigator, and boom operator, who is responsible for air refueling), we had discussed the fact that we would most likely land on one of the land-based 22 runways, as there was a westerly wind that day. Sure enough, we found ourselves looping Diamond Head and Honolulu proper to land to the west. However, surprisingly we were cleared to land on the reef Runway 26L, which was normally reserved for takeoffs. Once cleared for the visual approach, the aircraft commander aggressively lined up on Runway 22L (the wrong runway). I said, "Ace, we are cleared to Runway 26L." He replied, "Roger that." We continued toward 22L. I said again, "Ace, we are cleared 26L, correct?" He said, "I know that." I said, "We aren't lined up very well." He said, "It isn't too bad." He was thinking it was a pretty good approach and that I was a perfectionist. I was thinking, "It's going to be hard to land on 26L from here." Finally, our old, crusty boom operator quickly unstrapped, got out of his seat, and leaped forward to put his head right over the aircraft commander's shoulder. "IT'S THE REEF RUNWAY!" he said as he pointed emphatically. "Oh...the reef runway," said the AC as he realized that is 26L and quickly slid over (we were on about a 2 mi final at the time).

Ready on the set

How could the AC allow himself to line up on the wrong runway, 22L, after he clearly heard, and the copilot acknowledged, a landing on 26L? There could be a number of reasons, including fatigue and/or information overload. I think in this case there is a more parsimonious explanation. But first, read the following paragraph and count the number of times you see the letter F:

> FINISHED FIXTURES ARE THE PRODUCT OF YEARS OF SCIENTIFIC STUDIES COMBINED WITH THE EXPERIENCE OF NUMEROUS YEARS.

How many did you count? Was the number three? That is what I came up with the first time. If that is what you got, try it again. Did you find more this time? Are there three, four, or five? Actually there are six. Six, that seems hard to believe for many people. People frequently come up with less than that (three is the most common response), but as you look, there are indeed six. Did you notice the word *of*? It occurs three times, as well as the words *finished, fixtures,* and *scientific,* for a total of six Fs. Why would someone educated enough to read at a high level not be able to find the simple letter F? The answer is perceptual sets (Hughes et al. 1999). Perceptual sets are mind frames that we get into. (We don't expect the letter F to make a "v" sound. Yet it does in the word *of,* so our mind has a different expectation, and we overlook the letter F.) Perceptual sets get us into a certain frame of mind with certain expectations. We may be fully expecting event A, but event B occurs instead. When event B occurs, it is very surprising. It catches us totally off guard. I've had this occur at dinner when I reach for my Sprite, only to find upon tasting it that I have my water.

In this situation, we as a crew, and in particular as a pilot team, had rehearsed our expectation that we would almost certainly be landing on Runway 22L. This was standard operating procedure at Honolulu. "Expect 22L on arrival" was called over the ATIS, and we had both flown into Honolulu previously and found this to be the case. In the AC's mind, 22 Left is where we were going. When the controller gave us 26L, that was 22L in the AC's mind. Notice also that 22 and 26 are very similar numbers. Add to that they are both left runways. There is a Runway 22R and a Runway 26R as well to add to the confusion. So 22L is not very different from 26L. Often it requires outside intervention, whether another human or an automated backup system, to shake us out of our perceptual sets. They are that powerful.

San Diego—What a Great City

It was day VFR, a nice weekday morning. Bill Phillips was the captain of the large passenger airliner, with over 10,000 h under his belt. San Diego, considered by many to be the most beautiful city in the United States and among the most desirable to live in, was the destination. The aircraft was cleared the localizer Runway 27 approach into San Diego International Lindbergh Field (SAN). The DME was not reliable, and this information was on the ATIS, but the crew missed it. The DME appeared good—steady mileage readout, no flags. The crew descended below specified altitudes on the approach—specifically, 2000 ft to REEBO—the initial approach point (IAP). Approach got a low-altitude alert on the aircraft and asked the crew if it was in visual conditions. It was, and Approach then cleared the crew for the visual approach to Runway 27. The large jet made an uneventful landing. (ASRS 421124)

The passenger syndrome

Like many of the ASRS reports, in ASRS 421124 the reporting official offered his own diagnosis. One of the items mentioned by the captain is that when "flying with an experienced and proficient crew member, you fail to cross-check as diligently as usual." Again, we see the complacency factor, and, again, it is related to CRM. A crew that uses all its resources wisely does not fail to cross-check one another, even when a crew member is experienced. This is easier said than done; we all have a natural tendency to let down our guard a bit when flying with an experienced crew member. It is a very natural thing to do. But it can also be dangerous.

Anecdotally, it seems that most of the accidents I have read about in the Air Force have occurred with either a fairly new pilot or a very experienced pilot. Military accident reports are full of accidents that have occurred with instructor pilots aboard the aircraft. Fight this tendency to get complacent.

Both of these pilots seem to be highly experienced, making them susceptible to a closely related phenomenon known as the *passenger syndrome*. This is the idea that when flying with a very experienced crew member, you are just along for the ride. If actually flying, this person is not too concerned with mistakes because "he or she will bail me out," or "Big Brother will bail me out." This situation can be compounded if the less-experienced pilot is very new; he or she may look at the more experienced pilot as Mom or Dad. The remedy for this problem is to fly as if you are the most experienced person on board. Assume the other person is one of the three blind mice and won't catch anything. Fly as if you are calling the shots, and you will avoid the passenger syndrome.

Another effective strategy to combat the passenger syndrome is offered by Richard Jensen. He recommends the "what if" technique. This is actually a little game you can play with yourself, where instead of daydreaming off into space, you invent some scenario and a solution. An example would be flying along at cruise altitude 45 min from your destination, when you ask yourself, *What if my destination airport closed at this point? What would I do?*

Our discussion of complacency is nearing the beating-on-a-dead-horse stage. So let me just say this: We have now reviewed three studies where complacency was the culprit in causing poor situational awareness. First, we had a crew busting an approach altitude and heading toward the ground because it was complacent in flying an instrument approach with no approach plate. Then we had a crew trying to land on the wrong runway because of complacency. Now we have another crew busting approach altitudes because of complacency in cross-checking one another. Complacency and situational awareness are intertwined.

RTFQ...LTTWT

As a teacher, after giving back a graded test, I often hear students say, "But sir, I thought you were asking A" (A being something I wasn't asking). Then I ask them, "What does the question say?" When they read the question, they then enter what I call the "stammering stage." "But sir, I thought it said this when I looked at it the first time." "But clearly that is not what it is asking; you need to RTFQ" is my reply. RTFQ stands for Read the Full Question. If you don't completely read the question, you may miss the essence of it. This is a lesson I'm sure each of has learned from the school of hard knocks.

There is a corollary to RTFQ in the flying world. In his diagnosis of the botched approach at San Diego, the captain makes this comment: "ATIS at a lot of airports has so much extra airport information that when using ACARS [Aircraft Communications Addressing and Reporting System; a data link allowing crews access to ATIS and other information, and to communicate with their company, almost like an onboard e-mail system], you tend to 'skim' through it." The captain is absolutely right in one sense. ATIS does have a lot of extra airport information that is tempting to skip over. Many times a lot of the information does not really apply to your particular flight, but sometimes it does. That is why it is on there; you need to hear it. Different aircraft need to hear different information, so the airport is forced to put it all on ATIS. The captain already mentioned that he was fly-ing with an experienced crew member. Surely, with the confidence of not having to watch the first officer like a hawk, one of the pilots could have taken time to listen to the complete ATIS report. It is interesting that the con-troller cleared the crew for the localizer approach with an unreliable DME. But it is also a bit humorous that the captain goes on to report, "ATC should have mentioned the DME 'unreliable' when giving the approach clear-ance." If you call ATC asking for the approach and tell it you have information Yankee from the ATIS, then it defeats the purpose of ATIS for the controller to remind you of what is on the ATIS report. That is the purpose of ATIS—to cut down on radio traffic. When you listen to ATIS it is your responsibility to get the information. It is not the controller's job to hold your hand to make sure you listened to it all. So the corollary to RTFQ is LTTWT: Listen to the Whole Thing. It could save you!

Controllers Are People, Too

"My name is Tim. I am a certified flight instructor (CFI) on the BE36. I was out on a 'hump day,' a nice Wednesday morning for the late fall. It was mixed IFR and VFR weather. I was in and out of clouds and had the Sacramento Valley and American River in sight from time to time. I was out with another pilot for some personal training and proficiency. I was on an IFR flight plan to Sacramento Executive and, following my last instrument approach there, I went missed approach for training. Sacramento Departure gave us a vector of 140 degrees from Sacramento VOR at 2500 ft MSL while awaiting clear-ance to return IFR to Buchanan Field in Concord, California (CCR). We followed the vector given us. Soon we realized this was a direct collision course with an antenna array (a TV tower farm) located approximately 5 to 10 mi southeast of Sacramento VOR! It was barely VFR conditions. We spotted the antennas approximately 1 mi out and changed course to avoid collision. Had it been actual IMC, the vectors ATC had given us could have taken us into direct contact with the antennas. The controller was apparently in training, as his supervisor would get on the radio and correct any radio traffic mistakes made. While we were still flying, the controller also vectored one other plane 140 degrees at 2000 ft MSL. This would be a very dangerous vector in actual IMC! I have since refused radar vectors southeast of Sacramento VOR, stating 'unable to comply due to obstruction.' Controllers seem unaware of the antenna obstruction. On the sectional, antennas are indicated to be 2000 ft AGL." (ASRS 423218)

One more lesson

This report teaches us yet one more lesson on complacency. We can't just blindly follow the controller. We must be vigilant, not complacent. Controllers are people, too. They make mistakes, they go through training, they learn on the job, and they have stress in and out of work that affects their performance. Just as Tim was increasing his proficiency, so was the controller. Tim was fortunate to have mixed VMC and IMC. He was able to spot the antennas and take evasive action. But what is a pilot to do under IFR conditions? Is he or she under the complete whim and fancy of the controller?

Granted, flying in IMC conditions does put us at the mercy of the controller to a degree. One thing a pilot can do is to become familiar with the charts to identify local obstructions. These obstructions can be found on a sectional and often on approach plates. A good technique is to "chum" an older chart to update it for the most recent changes in obstacles. This is particularly helpful if you are flying to a strange or unfamiliar field.

Tim and his counterpart on the flight did not choose to be sheep led to pasture by the controller. They teach us a good lesson. They also researched the antennas further on their sectionals after the incident, so they were prepared for the next flight into Sacramento Executive. What is not clear is whether they called the ATC facility to discuss the obstacles with the controllers. It seems clear from subsequent flights that the controllers are continuing to use this same bad vector, apparently unaware or unconcerned with the obstacle. Tim could help remedy this with a phone call or, if need be, a series of follow-up calls until he gets somebody's attention.

In the channel

Thus far we have clearly established the relationship of complacency and a lack of situational awareness. There is also another connection that needs to be highlighted: the relationship between channelized attention and lost situational awareness. We saw an example of this with Bob in Chap. 2. Recall Bob was taking off from Bellingham, Washington, and heading across the Puget Sound. He encountered bad weather and immediately returned to the field. But while fixating his attention on finding the field, he stopped flying the aircraft and clipped the top of a tree. To review, channelized attention is a condition where we focus on one feature of our environment to the exclusion of other important cues. These other cues are trying to tell us important things (such as "I am a tree; you are going to hit me").

Channelized attention and lost situational awareness go hand in hand, just as complacency and SA do. As mentioned at the beginning of this chapter, from 1989 to 1995 channelized attention leading to lost situational awareness was the number one contributing factor to Air Force mishaps. Perhaps the quintessential example of this phenomenon is not a CFIT accident in the truest sense of the term, but it bears mentioning as a powerful illustration of lost situational awareness caused by channelized attention. It is also included in its entirety because of its historical significance and the role it has played in aircrew training since its occurrence. Although this accident may not have been controlled flight into terrain, had SA not been lost by the crew, the flight could have been controlled flight onto the runway. The following excerpt is taken directly from the National Transportation Safety Board's report on the incident dated June 7, 1979. The only addition is my inclusion of time from holding (TFH) to give the reader a greater appreciation of the timeline.

United Airlines Flight 173, DC-8, at Portland, Oregon

"On December 28, 1978, United Airlines, Inc., Flight 173, a McDonnell-Douglas DC-8-61 (N8082U), was a scheduled flight from John F. Kennedy International Airport, New York, to Portland International Airport, Portland, Oregon, with an en route stop at Denver, Colorado.

"Flight 173 departed from Denver about 1447 PST (Pacific Standard Time) with 189 persons on board, including 6 infants and 8 crewmembers. The flight was cleared to Portland on an instrument flight rules (IFR) flight plan. The planned time en route was 2 h 26 min. The planned arrival time at Portland was 1713 PST.

"According to the automatic flight plan and monitoring systems the total amount of fuel required for the flight to Portland was 31,900 lb. There was 46,700 lb of fuel on board the aircraft when it departed the gate at Denver. This fuel included the Federal Aviation Regulation requirement for fuel to destination plus 45 min and the company contingency fuel of about 20 min. During a postaccident interview, the captain stated that he was very close to his predicted fuel for the entire flight to Portland '...or there would have been some discussion of it.' The captain also explained that his flight from Denver to Portland was normal.

"At 1705:47, Flight 173 called Portland Approach and advised that its altitude was 10,000 ft MSL (all altitudes in MSL unless otherwise noted) and its airspeed was being reduced. Portland responded and told the flight to maintain its heading for a visual approach to runway 28. Flight 173 acknowledged the approach instructions and stated, '...we have the field in sight.'

"At 1707:55, Portland Approach instructed the flight to descend and maintain 8,000 ft. Flight 173 acknowledged the instructions and advised that it was 'leaving ten.' At 1709:40, Flight 173 received and acknowledged a clearance to continue its descent to 6,000 ft.

"During a postaccident interview, the captain stated that, when Flight 173 was descending through about 8,000 ft, the first officer, who was flying the aircraft, requested the wing flaps be extended to 15, then asked that the landing gear be lowered. The captain stated that he complied with both requests. However, he further stated that, as the landing gear extended, "...it was noticeably unusual and [I] feel it seemed to go down more rapidly. As [it is] my recollection, it was a thump, thump in sound and feel. I don't recall getting the red and transient gear door light. The thump was much out of the ordinary for this airplane. It was noticeably different and we got the nose gear green light but no other lights.' The captain also said the first officer remarked that the aircraft 'yawed to the right....' Flight attendant and passenger statements also indicate that there was a loud noise and a severe jolt when the landing gear was lowered.

"At 1712:20, Portland Approach requested, 'United one seven three heavy, contact the tower [Portland], one one eight point seven.' The flight responded, 'Negative, we'll stay with you. We'll stay at five. We'll maintain about a hundred and seventy knots. We got a gear problem. We'll let you know.' This was the first indication to anyone on the ground that Flight 173 had a problem. At 1712:28, Portland Approach replied, 'United one seventy three heavy roger, maintain five thousand. Turn left heading two zero zero.' The flight acknowledged the instructions. At 1714:43 [time from holding (TFH) = 2 min], Portland Approach advised, 'United one seventy three heavy, turn left heading, one zero zero and I'll just orbit you out there 'til you get your problem.' Flight 173 acknowledged the instructions.

"For the next 23 minutes, while Portland Approach was vectoring the aircraft in a holding pattern south and east of the airport, the flightcrew discussed and accomplished all of the emergency and precautionary actions available to them to assure themselves that all landing gear was locked in the full down position. The second officer checked the visual indicators on top of both wings, which extend above the wing surface when the landing gear is down-and-locked.

"The captain stated that during this same time period, the first flight attendant came forward and he discussed the situation with her. He told her that after they ran a few more checks, he would let her know what he intended to do.

About 1738 [TFH = 26 min], Flight 173 contacted the United Airlines Systems Line Maintenance Control Center in San Francisco, California, through Aeronautical Radio, Inc. (Aeronautical Radio, Inc., an air-to-ground radio service which provides a communication system for commercial aircraft). According to recordings, at 1740:47

[TFH = 28 min] the captain explained to company dispatch and maintenance personnel the landing gear problem and what the flight crew had done to assure that the landing gear was fully extended. He reported about 7,000 lb of fuel on board and stated his intention to hold for another 15 or 20 minutes. He stated that he was going to have the flight attendants prepare the passengers for emergency evacuation.

"At 1744:03 [TFH = 32 min], United San Francisco asked, 'Okay, United one seventy three, you estimate that you'll make a landing about five minutes past the hour. Is that okay?' The captain responded, 'Ya, that's good ballpark. I'm not gonna hurry the girls. We got about a hundred-sixty-five people on board and we...want to...take our time and get everybody ready and then we'll go. It's clear as a bell and no problem.'

"The aircraft continued to circle under the direction of Portland Approach in a triangular pattern southeast of the airport at 5,000 ft. The pattern kept that aircraft within about 20 nautical miles of the airport.

"From about 1744:30 until about 1745:23 [TFH = 33 min], the cockpit voice recorder (CVR) contained conversation between the captain and the first flight attendant concerning passenger preparation, crash landing procedures, and evacuation procedures. During his initial interview, the captain indicated that he neither designated a time limit to the flight attendant nor asked her how long it would take to prepare the cabin. He stated that he assumed 10 or 15 minutes would be reasonable and that some preparations could be made on the final approach to the airport.

"At 1746:52 [TFH = 34 min], the first officer asked the flight engineer, 'How much fuel we got...?' The flight engineer responded, 'Five thousand.' The first officer acknowledged the response.

"At 1748:38 [TFH = 36 min], Portland Approach advised Flight 173 that there was another aircraft in its vicinity. The first officer advised Portland Approach that he had the aircraft in sight.

"At 1748:54 [TFH = 36 min], the first officer asked the captain, 'What's the fuel show now...?' The captain replied, 'Five.' The first officer repeated, 'Five.' At 1749 [TFH = 37 min], after a partially unintelligible comment by the flight engineer concerning fuel pump lights, the captain stated, 'That's about right, the feed pumps are starting to blink.' According to data received from the manufacturer, the total usable fuel remaining when the inboard feed pump lights illuminate is 5,000 lb. At this time, according to flight data recorder (FDR) and air traffic control data, the aircraft was about 13 nautical miles south of the airport on a west southwesterly heading.

"From just after 1749 until 1749:45 [TFH = 37 min], the flightcrew engaged in further conversation about the status of the landing gear. This conversation was interrupted by a heading change from Portland Approach and was followed by a traffic advisory from Portland Approach.

"About 1750:20 [TFH = 38 min], the captain asked the flight engineer to 'Give us a current card on weight. Figure about another fifteen minutes.' The first officer responded, 'Fifteen minutes?' To which the captain replied, 'Yeah, give us three or four thousand pounds on top of zero fuel weight.' The flight engineer then said, 'Not enough. Fifteen minutes is gonna—really run us low on fuel here.' At 1750:47 [TFH = 38 min], the flight engineer gave the following information for the landing data card: 'Okay. Take three thousand pounds, two hundred and four.' At this time the aircraft was about 18 nautical miles south of the airport in a turn to the northeast.

"At 1751:35 [TFH = 39 min], the captain instructed the flight engineer to contact the company representative at

Portland and apprise him of the situation and tell him that Flight 173 would land with about 4,000 lb of fuel. From 1752:17 until about 1753:30 [TFH = 41 min], the flight engineer talked to Portland and discussed the aircraft's fuel state, the number of persons on board the aircraft, and the emergency landing preparations at the airport. At 1753:30 [TFH = 41 min], because of an inquiry from the company representative at Portland, the flight engineer told the captain, 'He wants to know if we'll be landing about five after.' The captain replied, 'Yea.' The flight engineer relayed the captain's reply to the company representative. At this time the aircraft was about 17 nautical miles south of the airport heading northeast.

"At 1755:04 [TFH = 43 min], the flight engineer reported the '...approach descent check is complete.' At 1756:53 [TFH = 44 min], the first officer asked, 'How much fuel you got now?' The flight engineer responded that 4,000 lb remained, 1,000 lb in each tank.

"At 1757:21 [TFH = 45 min], the captain sent the flight engineer to the cabin to 'kinda see how things are going.' From 1757:30 until 1800:50 [TFH = 48 min], the captain and the first officer engaged in a conversation which included discussions of giving the flight attendants ample time to prepare for the emergency, cockpit procedures in the event of an evacuation after landing, whether the brakes would have antiskid protection after landing, and the procedures the captain would be using during the approach and landing.

"At 1801:12 [TFH = 49 min], Portland Approach requested that the flight turn left to a heading of 195 deg. The first officer acknowledged and compiled with the request.

"At 1801:34, the flight engineer returned to the cockpit and reported that the cabin would be ready in 'another two or three minutes.' The aircraft was about 5 nautical miles southeast of the airport turning to a southwesterly heading. Until about 1802:10 [TFH = 50 min], the captain and the flight engineer discussed the passengers and their attitudes toward the emergency.

"At 1802:22 [TFH = 50 min], the flight engineer advised, 'We got about three on the fuel and that's it.' The aircraft was then about 5 nautical miles south of the airport on a southwest heading. The captain responded, 'Okay. On touchdown, if the gear folds or something really jumps the track, get those boost pumps off so that...you might even get the valves open.'

"At 1802:44, Portland Approach asked Flight 173 for a status report. The first officer replied, 'Yeah, we have indication our gear is abnormal. It'll be our intention, in about five minutes, to land on two eight left. We would like the equipment standing by. Our indications are the gear is down and locked. We've got our people prepared for an evacuation in the event that should become necessary.'

At 1803:14 [TFH = 51 min], Portland Approach asked that Flight 173 advise them when the approach would begin. The captain responded, 'They've about finished in the cabin. I'd guess about another three, four, five minutes.' At this time the aircraft was about 8 nautical miles south of the airport on a southwesterly heading.

"At 1803:23 [TFH = 51 min], Portland Approach asked Flight 173 for the number of persons on board and the amount of fuel remaining. The captain replied, '...about four thousand, well, make it three thousand, pounds of fuel,' and 'you can add to that one seventy-two plus six laps-infants.'

"From 1803:38 until 1806:10 [TFH = 54 min], the flight crew engaged in a conversation which concerned (1) checking the landing gear warning horn as further evidence that the landing gear was fully down and locked and (2) whether automatic spoilers and antiskid would operate normally with the landing gear circuit breakers out.

"At 1806:19 [TFH = 54 min], the first flight attendant entered the cockpit. The captain asked, 'How you doing?' She responded, 'Well, I think we're ready.' At this time the aircraft was about 17 nautical miles south of the airport on a southwesterly heading. The conversation between the first flight attendant and the captain continued until about 1806:40 [TFH = 54 min] when the captain said, 'Okay. We're going to go in now. We should be landing in about five minutes.' Almost simultaneous with this comment, the first officer said, 'I think you just lost number four...' followed immediately by advice to the flight engineer, '...better get some crossfeeds open there or something.'

"At 1806:46 [TFH = 54 min], the first officer told the captain, 'We're going to lose an engine...' The captain replied, 'Why?' At 1806:49, the first officer again stated, 'We're losing an engine.' Again the captain asked, 'Why?' The first officer responded, 'Fuel.'

"Between 1806:52 and 1807:06 [TFH = 55 min], the CVR revealed conflicting and confusing conversation between flight crewmembers as to the aircraft's fuel state. At 1807:06, the first officer said, 'It's flamed out.'

"At 1807:12, the captain called Portland Approach and requested, '...would like clearance for an approach into two eight left, now.' The aircraft was about 19 nautical miles south southwest of the airport and turning left. This was the first request for an approach clearance from Flight 173 since the landing gear problem began. Portland Approach immediately gave the flight vectors for a visual approach to runway 28L. The flight turned toward the vector heading of 010 degrees.

"From 1807:27 until 1809:16 [TFH = 57 min], the following intracockpit conversation took place:

1807:27—Flight Engineer 'We're going to lose number three in a minute, too.'

1807:31—Flight Engineer 'It's showing zero.'

Captain 'You got a thousand pounds. You got to.'

Flight Engineer 'Five thousand in there...but we lost it.'

Captain 'Alright.'

1807:38—Flight Engineer 'Are you getting it back?'

1807:40—First Officer 'No number four. You got that crossfeed open?'

1807:41—Flight Engineer 'No, I haven't got it open. Which one?'

1807:42—Captain 'Open 'em both—get some fuel in there. Got some fuel pressure?'

Flight Engineer 'Yes sir.'

1807:48—Captain 'Rotation. Now she's coming.'

1807:52—Captain 'Okay, watch one and two. We're showing down to zero or a thousand.'

Flight Engineer 'Yeah.'

Captain 'On number one?'

Flight Engineer 'Right.'

1808:08—First Officer 'Still not getting it.'

1808:11—Captain 'Well, open all four crossfeeds.'

Flight Engineer 'All four?'

Captain 'Yeah.'

1808:14—First Officer 'Alright, now it's coming.'

1808:19—First Officer 'It's going to be—on approach, though.'

Unknown Voice 'Yeah.'

1808:42—Captain 'You gotta keep 'em running....'

Flight Engineer 'Yes, sir.'

1808:45—First Officer 'Get this...on the ground.'

Flight Engineer 'Yeah. It's showing not very much more fuel.'

1809:16—Flight Engineer 'We're down to one on the totalizer. Number two is empty.'"

"At 1809:21 [TFH = 57 min], the captain advised Portland Approach, 'United seven three is going to turn toward the airport and come on in.' After confirming Flight 173's intentions, Portland Approach cleared the flight for the visual approach to Runway 28L.

"At 1810:17 [TFH = 58 min], the captain requested that the flight engineer 'reset that circuit breaker momentarily. See if we get gear lights.' The flight engineer complied with the request.

"At 1810:47, the captain requested the flight's distance from the airport. Portland approach responded, 'I'd call it eighteen flying miles.' At 1812:42 [TFH = 60 min], the captain made another request for distance. Portland Approach responded, 'Twelve flying miles.' The flight was then cleared to contact Portland tower.

"At 1813:21 [TFH = 61 min], the flight engineer stated, 'We've lost two engines, guys.' At 1813:25, he stated, 'We just lost two engines—one and two.'

"At 1813:38, the captain said, 'They're all going. We can't make Troutdale' (a small airport on the final approach path to runway 28L). The first officer said, 'We can't make anything.'

"At 1813:46, the captain told the first officer, 'Okay. Declare a mayday.' At 1813:50, the first officer called Portland International Airport tower and declared, 'Portland tower, United one seventy three heavy, Mayday. We're—the engines are flaming out. We're going down. We're not going to be able to make the airport.' This was the last radio transmission from Flight 173.

"About 1815 [TFH = 63 min], the aircraft crashed into a wooded section of a populated area of suburban Portland about 6 nautical miles east southeast of the airport. There was no fire. The wreckage path was about 1,554 ft long and about 130 ft wide."

The dawn of CRM

What happened here? In a nutshell, the captain and the rest of the crew lost SA because of channelized attention on a landing gear problem. Of the 181 passengers and 8 crew members aboard, 8 passengers, the flight engineer, and a flight attendant were killed and 21 passengers and 2 crew members were injured seriously. The aircraft smashed into a wooded area and two houses upon crash landing. Fortunately, both houses were unoccupied at the time (though there were numerous occupied houses and apartments in the near vicinity). This accident produced a recommendation from the NTSB, which in turn spawned a movement called CRM. I use this case as a lead-in to CRM, which we cover in the next chapter.

Keep the main thing, the main thing

I have read several things on bathroom walls in my lifetime. Very few of them bear repeating. However, I once came upon something actually profound on one wall. It said simply, "The main thing is to keep the main thing, the main thing." That is very good advice. Unfortunately, this crew did not follow it.

To be clear, there was a problem with the landing gear. An investigation after the accident centered "on the piston rod and the mating end from the right main landing gear retract cylinder assembly." It was found that there was a "separation of the rod end from the piston rod due to severe corrosion caused by moisture on the mating threads of both components." "The corrosion allowed the two parts to pull apart and the right main landing gear to fall free when the flight crew lowered the landing gear. This rapid fall disabled the microswitch for the right main landing gear which completes an electrical circuit to the gear-position indicators in the cockpit. The difference between the time it took for the right main landing gear to free-fall and the time it took for the left main landing gear to extend normally probably created a difference in aerodynamic drag for a short time. This difference in drag produced a transient yaw as the landing gear dropped."

While it is clear that the landing gear did not drop normally, was this a cause for concern and would it keep the aircraft from landing? Not according to the NTSB: "If the visual indicators indicate the gear is down, then a landing can be made at the captain's discretion." The flight engineer's check of the visual indicators for both main landing gear showed that they were down and locked. A visual check of the nose landing gear could not be made because the light that would have illuminated that down-and-locked visual indicator was not operating. However, unlike the main landing gear cockpit indicators, the cockpit indicator for the nose gear gave the proper "green gear-down" indication.

Again, from the NTSB report: "Admittedly, the abnormal gear extension was cause for concern and a flightcrew should assess the situation before communicating with the dispatch or maintenance personnel. However, aside from the crew's discussing the problem and adhering to the DC-8 Flight Manual, the only remaining step was to contact company dispatch and line maintenance. From the time the captain informed Portland Approach of the gear problem until contact with company dispatch and line maintenance, about 28 min had elapsed. The irregular gear check procedures contained in their manual were brief, the weather was good, the area was void of heavy traffic, and there were no additional problems experienced by the flight that would have delayed the captain's communicating with the company. The company maintenance staff verified that everything possible had been done to assure the integrity of the landing gear.

Therefore, upon termination of communications with company dispatch and maintenance personnel, which was about 30 min before the crash, the captain could have made a landing attempt. The Safety Board believes that Flight 173 could have landed safely within 30 to 40 min after the landing gear malfunction."

It seems clear that the gear problem in and of itself was not sufficient to cause a crash landing. The other possibility is that preparation for crash landing forced the crew to hold for 1 h. The captain certainly did not leave the flight attendant any impression that time was of the essence; note the following from the NTSB: "Upon completing communications with company, line maintenance and dispatch, the captain called the first flight attendant to the cockpit to instruct her to prepare the cabin for a possible abnormal landing. During the ensuing discussion, the captain did not assign the first flight attendant a specified time within which to prepare the cabin, as required by the flight manual. In the absence of such time constraint, the first flight attendant was probably left with the impression that time efficiency was not necessarily as important as the assurance of thorough preparation." The report went on to say that the captain did not feel that the preparations of the flight attendants were driving the decision of when to land. "In the initial interview with the captain, he stated that he felt the cabin preparation could be completed in from 10 to 15 min and that the 'tail end of it' could be accomplished on the final approach to the airport."

Clearly, the problem lay with the captain and the rest of the crew failing to monitor the fuel situation adequately. Though at a number of points crew members discussed fuel remaining, at no time did they translate that into time remaining to fly given that fuel. Also, no one was given responsibility to monitor the fuel and keep the crew informed. "Therefore, the Safety Board can only conclude that the flightcrew failed to relate the fuel remaining and the rate of fuel flow to the time and distance from the airport, because their attention was directed almost entirely toward diagnosing the landing gear problem." Indeed, the crew seemed oblivious to its upcoming fate: "This failure to adhere to the estimated time of arrival and landing fuel loads strengthens the Board's belief that the landing gear problem had a seemingly disorganizing effect on the flightcrew's performance. Evidence indicates that their scan of the instruments probably narrowed as their thinking fixed on the gear. After the No. 4 engine had flamed out and with the fuel totalizer indicating 1,000 lb, the captain was still involved in resetting circuit breakers to recheck landing gear light indications." It seems that the crew was experiencing temporal distortion, which we discussed earlier. Time seemed to be standing still, but it wasn't.

Recently, there was a published interview with the controller on duty that night. His story adds an interesting twist to this ill-fated flight. "I was one of the Portland International FAA air traffic controllers on duty that fateful evening. I was working arrival radar south and took the handoff from Seattle Center on UA173. The flight was in a position for a straight-in ILS approach to Runway 28R, and I instructed the pilot to intercept the localizer and proceed inbound. At some point the crew advised me they were having a gear problem. I offered Captain X [name withheld to protect anonymity] the option of holding over the Laker outer compass locater at 6,000 until they were able to resolve the problem. (From that position, they could have deadsticked to either runway.) Captain X declined and canceled IFR. He elected to proceed southeast of the airport 20 miles or so, and circled for a period of time." So we see that even the controller was trying to put the aircrew in position to land in case of an immediate problem. Had the crew taken the controller's offer, it could have dead-sticked (flown without power) a landing even after running the aircraft out of fuel. Back to the NTSB findings: "Throughout the landing delay, Flight 173 remained at 5,000 ft with landing gear down and flaps set at 15 deg. Under these conditions, the Safety Board estimated that the flight would have been burning fuel at the rate of about 13,209 lb per hour—220 lb per min. At the beginning of the landing delay, there were about 13,334 lb of fuel on board." No crew member ever verbalized this fact. Who knows if it ever crossed anyone's mind?

The NTSB cited the following as the probable cause of the crash of UAL Flight 173. "The National Transportation Safety Board determined that the probable cause of the accident was the failure of the captain to monitor properly the aircraft's fuel state and to properly respond to the low fuel state and the crewmember's advisories regarding fuel state. This resulted in fuel exhaustion to all engines. His inattention resulted from preoccupation with a landing gear malfunction and preparations for a possible landing emergency. Contributing to the accident was the failure of the other two flight crewmembers either to fully comprehend the criticality of the fuel state or to successfully communicate their concern to the captain." An airline crew is not the only group susceptible to lost SA because of channelized attention. Morrison (1993) and a team of researchers found the following in a study of general aviation pilots. "Distractions and related attention failures are the key factors in many landing incidents and accidents." In sum, the crew of Flight 173 failed to keep the main thing, the main thing. Don't you make the same mistake.

Summary

In this chapter we have seen how lost situational awareness can result in CFIT or CFTT. Two concepts seem to be precursors to lost SA: complacency and channelized attention. Both can result in an insidious onset of lost situational awareness without the crew's knowledge. Chris Wickens, one of the United States' foremost experts on human factors and aviation psychology, draws two conclusions from research on situational awareness. The process of maintaining good situational awareness depends on two mental components: long-term memory and attention. Through long-term memory we are able to construct a framework to store the information of what has just happened, what is happening, and what is likely to happen. Selective attention is critical to direct the eyes and the mind to appropriate places to acquire the relevant information. It is difficult to ascertain what is happening if attention is not directed over the total—the big picture.

Wickens also points out that experts can be susceptible to lost SA just as a novice can (as we have seen from several of the case studies). There are three reasons this can occur:

1. High workload can direct the expert's attention to the wrong channels of information (e.g., the UAL 173 crew's obsession with gear checks).

2. Displays giving information that does not stand out to the pilot or giving information that is salient about only portions of the changing environment (e.g., the UAL 173 gear indications were clear, while the crew had an underappreciation for what the fuel gages and flowmeters were telling it).

3. Finally, experts have years of experience, which may lead them to perceptual sets or mindsets that in turn lead them to an incomplete sampling of

information; or when they get the information, they intercept it in an incorrect or biased way (the UAL 173 crew focused on 5000 lb of fuel rather than what amount of time that gave it to fly).

Wickens proposes two ways to improve this situation. First, crews, both novices and experts, can be trained in scanning skills to improve selective attention and to ensure they get the right information to their minds. Second, we can improve display technology and its ability to convey information to the pilot with an easier-to-understand picture of the dynamic flight environment. Proper display development can further improve the pilot's scanning ability and selective attention skills.

All of the instances of lost SA in this chapter could have been avoided with a proper attitude of vigilance among crew members. Vigilance embodies the idea of being on the alert and watchful. Think of the bald eagle; the powerful bird is never just bebopping along. Remember the eyes; the eyes are vigilant. Even as the body soars effortlessly on the wings, the eyes remain vigilant—always seeking, always thinking. Such should be our flying.

References and For Further Study

Close Up: United Airlines Flight 173, at
  http://www.avweb.com/articles/ual173.html.
Endsley, M. R. 1989. Pilot situation awareness: The challenge for the training community. Paper presented at the Interservice/Industry Training Systems Conference, Fort Worth, Tex. November.
Endsley, M. R. 1995. Toward a theory of situation awareness in dynamic systems. *Human Factors,* 37(1):85–104.

Gopher, D. 1992. The skill of attention control: Acquisition and execution of attention strategies. In *Attention and Performance XIV: Synergies in Experimental Psychology, Artificial Intelligence, and Cog Neuroscience—A Silver Jubilee.* Eds. D. E. Meyer and Kornblum. Cambridge, Mass.: MIT Press.

Gopher, D., M. Weil, and T. Bareket. 1994. Transfer of skill from a computer game trainer to flight. *Human Factors,* 36(4):387–405.

Hughes, R., R. Ginnett, and G. Curphy. 1999. *Leadership: Enhancing the Lessons of Experience,* 3d ed. Boston: Irwin/McGraw-Hill. Chap. 4.

Human Factors. 1995. Special Issue: Situation Awareness.

Jensen, R. 1995. *Pilot Judgment and Crew Resource Management.* Aldershot, England: Avebury Aviation.

Kahneman, D., P. Slovic, and A. Tversky. 1982. *Judgment under Uncertainty: Heuristics and Biases.* New York: Cambridge University Press.

Magnuson, K. 1995. Human factors in USAF mishaps, 1 Oct 89 to 1 Mar 95. USAF Safety Agency Life Sciences Division, Albuquerque, N.Mex.

Morrison, R., K. Etem, and B. Hicks. 1993. General aviation landing incidents and accidents: A review of ASRS and AOPA research findings. In *Proceedings of the Seventh International Symposium on Aviation Psychology,* 975–980.

Tversky, A. and D. Kahneman. 1974. Judgment under uncertainty: Heuristics and biases. *Science,* 185:1124–1131.

Wickens, C. 1999. Cognitive factors in aviation. In *Handbook of Applied Cognition.* Eds. F. T. Durso, R. S. Nickerson, R. W. Schvaneveldt, S. T. Dumais, D. S. Lindsay, and M. T. H. Chi. New York: Wiley. pp. 247–282.

5

Crew Resource Management

I concluded the last chapter with one of the most important case studies of all time in relationship to aircraft accidents and prevention. UAL Flight 173 into Portland, Oregon, in December of 1978 marked a turning point in aircrew training. The NTSB has published a work entitled, "We Are All Safer—NTSB-Inspired Improvements in Transportation Safety." In this publication, Flight 173 is discussed and its impact on crew resource management (CRM) is chronicled. Following is an excerpt.

Crew Resource Management

"In a number of airline accidents investigated by the Safety Board in the 1960s and 1970s, the Board detected a culture and work environment in the cockpit that, rather than facilitating safe transportation, may have contributed to the accidents. The Board found that some captains treated their fellow cockpit crewmembers as underlings who should speak only when spoken to. This intimidating atmosphere actually led to accidents when critical information was not communicated among cockpit crewmembers. A highly publicized accident in 1978 provided the impetus to change this situation.

"On December 28, 1978, as a result of a relatively minor landing gear problem, a United Airlines DC-8 was in a holding pattern while awaiting landing at Portland, Oregon. Although the first officer knew the aircraft was low on fuel, he failed to express his concerns convincingly to the captain. The plane ran out of fuel and crashed, killing 10.

"As a result of this accident and others, the concept of cockpit resource management, now called crew resource management (CRM), was born. Following pioneering work by the National Aeronautics and Space Administration (NASA), the Safety Board issued recommendations to the FAA and the airline industry to adopt methods that encourage teamwork, with the captain as the leader who relies on the other crewmembers for vital safety-of-flight tasks and also shares duties and solicits information and help from other crewmembers. United Airlines was one of the first airlines to adopt this concept, which is endorsed by pilot unions and is now almost universally used by the major airlines (as well as in other modes of transportation). The Board has also recommended and the FAA has acted to implement CRM for regional and commuter airlines.

"The value of CRM was demonstrated on July 19, 1989, when a United Airlines DC-10 experienced a catastrophic engine failure over Iowa that destroyed the aircraft's hydraulic systems, rendering it virtually uncontrollable. The cockpit crew and a deadheading captain who was a passenger worked as a team to bring the aircraft down to a crash landing at Sioux City. Although more than 100 people perished, almost 200 survived a situation for which no pilots in the world had ever been trained."

Defining CRM

It is clear from the preceding that Flight 173 was the impetus for mandated CRM training among both commercial and regional airlines. The following safety recommendation came from the NTSB report on Flight 173: "Issue an operations bulletin to all air carrier operations inspectors directing them to urge their assigned operators to ensure that their flightcrews are indoctrinated in principles of flightdeck resource management, with particular emphasis on the merits of participative management for captains and assertiveness training for other cockpit crewmembers." This *flight deck resource management* is what we now know as CRM. I will devote a lot of space to CRM as it is foundational to preventing CFIT in crew and fighter aircraft.

Tony Kern defines crew resource management as follows: "CRM means maximizing mission effectiveness and safety through effective utilization of all available resources." Though definitions abound and the term is thrown around loosely, that is the best definition I have seen: effectively utilizing all available resources—those in the air as well as on the ground. Lest the single-seat fighter think he is exempt, I would point out that effective use of your wingmen clearly falls under the realm of CRM.

Let's quickly review UAL Flight 173 into Portland, this time with an eye toward CRM. It is clear that the aircraft had a minor gear malfunction that needed to be dealt with. The crew also had a responsibility to prepare the passengers for a possible crash landing. While accomplishing these two tasks, the crew failed to follow the three basic rules of an aircraft emergency:

1. Maintain aircraft control.

2. Analyze the situation and take proper action.

3. Land as soon as conditions permit.

The crew violated Rule 1 by allowing the aircraft to get into a situation where aircraft control could not be maintained. With Rule 2, the crew analyzed a portion of the situation but failed to take into account the entire situation and to act properly. The crew discussed the fuel level but never translated that fuel level into an estimate of possible flight time. It acted as though the fuel would never run out but would simply just get lower. The captain failed to assign someone to act as the "timekeeper" and never calculated a bingo fuel level and a bingo position at which the crew would commit to returning to Portland International for landing. In a nutshell, the crew members failed to use all resources for the requisite variety of required duties and rather committed all of their human resources to the landing gear problem and checking on the status of the passengers. Finally, they violated Rule 3 by not landing as soon as conditions permitted. The NTSB accident report estimated that the crew could have landed anywhere from 23 to 33 min before the crash with no compromise of safety.

Though not clearly stated (but inferred), in the accident report and in subsequent publications is a possibly negative personality trait of the captain. It seems that the captain was a bit overbearing with other crew members. It has been said that he was not open to input from fellow crew members and liked to call the shots. Some evidence for this claim comes from the fact that the first and second officers sent some veiled messages to the captain concerning the fuel state. They did not dissuade the captain in the least as he continued down the trail of disaster. Perhaps he wasn't able to interpret the messages or perhaps he chose to ignore them. It's hard to say. Whatever the actual situation, the accident board found that "the first officer's and the flight engineer's inputs on the flight deck are important because they provide redundancy. The Safety Board believes that, in training of all airline cockpit and cabin crewmembers, assertiveness training should be a part of the standard curricula, including the need for individual initiative and effective expression of concern." This expression can start out veiled (depending on time constraints) but should not remain that way if the subtle hints do not change behavior and actions. The other crew members must be forthright.

The captain is clearly responsible for the final decisions aboard the aircraft. Without this authority, the cockpit is open ground for anarchy—not a desirable environment for any crew member or passenger. While responsibility and willingness to make a decision are admirable traits, the captain who stifles input from others in the cockpit is a walking time bomb. No one is perfect and everyone can benefit from knowledgeable input from other professionals. It is no shame to see things in a new light because your first officer is willing to speak up. As we will see in our next scenario, the first officer's willingness to speak up can save the day. Throwing hints that aren't taken doesn't do the job.

Viking Country

"We were in Viking country, the most beautiful time of year, August. The warm days led to comfortable evenings. The black flies were gone; it is the best time to be in Minnesota. It was getting to be dusk. We were in our SAAB 340, our two turboprop engines churning along. The SAAB is a two-crew transport, and we were approaching International Wold-Chamberlain with passengers having departed from Minneapolis/St. Paul. This would be a full stop for the last flight of the day. We were within 30 nautical mi (NM) of the airport and ready to call it a day.

"After being cleared for the approach for Runway 30L, Air Traffic Control advised a thunderstorm cell was on the field and he had received a low-level wind shear and microburst alert. The area was a mix of IMC/VMC. At this point we were in visual conditions, under the scattered-broken layer, and could see the leading edge of the cell and its rain shaft. I was the first officer with over 2200 h flight time and 450 h in this bird. I felt comfortable with my position, so I stated to the captain, 'This is a textbook microburst scenario.' He (the pilot flying) continued the approach. Another commuter aircraft, Y, told Tower it was 50/50 whether it would continue the approach and it would advise. Since misery loves company, I queried Air Traffic Control as to the location of Aircraft Y. Air Traffic Control said Aircraft Y was 3 mi behind our aircraft in trail. I said to the captain, 'Maybe we should go around.' The captain responded, 'No, let's keep going.' So we continued the approach.

"A few moments later, Air Traffic Control advised 'microburst alert' and wind shear reported by aircraft on the ground. The pilot blurted out, 'When was that?' I queried ATC, 'When was that wind shear reported by ground aircraft?' Air Traffic Control replied '...current...' The pilot flying began to fidget, saying '...I'm not sure...' I sensed this was my chance to again register my input, so I said, 'Well, let's go around.' To my astonishment, the pilot flying made no comment and continued the approach. Subsequently, I said nothing and we configured the aircraft for landing. As we continued on the approach the pilot flying said, 'I'm going to keep it fast," referring to our airspeed. At approximately 500 ft AGL, wham!, we encountered heavy rain, as we entered the leading edge of the cell.

"As the pilot not flying, I was reporting airspeed and normal approach calls. We noticed no large flux in speed but did notice a very strong crosswind from approximately 040–070 degrees and excessive crab angle to the right of about 45 degrees at 250–200 ft AGL. Indicated airspeed was relatively steady at 130 knots, required crab was increasing, and the aircraft was drifting left of centerline. Simultaneously, the pilot flying and I shouted, 'Go around!' The captain, still at the controls, barked, 'Set max power, flaps 7 degrees.' Again, I immediately complied— 'Power set...positive rate'; we were flying away. The pilot flying said, 'Call Air Traffic Control going around...gear up flaps up after takeoff checklist.'

"The worst was over, or so I thought. Air Traffic Control gave us a right-hand turn to the south. A few moments later the controller asked, 'Would you like a visual to Runway 04?'" I glanced at the captain and confirmed, 'Yes, visual to Runway 04.' I looked up, and to my horror, the right-turn bank exceeded 45 degrees. 'Watch your bank!' I screamed. The pilot flying corrected but not quickly enough. We descended and received a 'terrain' and 'whoop whoop pull up' warning. The GPWS was squawking. The captain recovered by climbing aggressively, adding power. Vigilance was high at this point, to say the least. We continued toward a normal visual approach procedure and landing on Runway 04. Again we encountered a strong crosswind from approximately 90 degrees; however, the crosswind was not as strong as the one we encountered on Runway 30L. This portion of the approach and the subsequent landing were fortunately uneventful. Once we were on the ground, my mind traveled back to the initial crew meeting before the first flight of the day. In an informative portion of the meeting, the pilot mentioned that he had gotten only 3 h of sleep the night before." (ASRS 410887)

The moral courage to speak up

This aircrew had several factors to deal with, not the least of which was the bad weather. This subject has already been thoroughly covered in Chap. 2. Recall from the freighter going into Belize that there is a good option when approaching a field experiencing storms and/or wind shear. You should consider heading to the holding pattern and working on your timing on each leg until the storm blows through. "Get-home-itis" can get you. Save yourself the trouble and spend some time in holding (fuel permitting) to ensure a safe landing. Also remember to use any weather radar on board or contact the weather station on the ground to ascertain how long the storms may stick around.

Fatigue is another factor that jumps out in this case. As discussed earlier, fatigue can make us complacent. Furthermore, fatigue slows our mental processes. We aren't as sharp or quick when we are fatigued. The captain had only 3 h of sleep the night before. Most experts recommend 12 h of crew rest, with 8 of those hours being spent "horizontal," fast asleep. Three hours at night is not enough time to refresh any non-nocturnal creature out there. That is likely why the captain did not respond when the first officer suggested they go around after the captain had expressed uncertainty.

Fatigue impairs our judgment. That may have been why the captain initially seemed to underestimate the weather. Later, the experienced first officer was bold enough to suggest a go-around when the commuter in trail stated that it was 50/50 about completing the approach. Even with these factors the captain shrugged them off with, "No, let's keep going." With the captain in this condition, the first officer was right on when he finally called "Go around!" with the aircraft steadily drifting left of centerline. Fortunately, the captain arrived at the same conclusion simultaneously. That was the right decision.

Yet the first officer continued his good cross-check of the captain. It was the first officer who caught the excessive bank and subsequent descent on the part of the tired captain. The automated GPWS (Ground Proximity Warning System) also kicked in to help the worsening situation. Perhaps because of the forewarning by the captain on his sleep deprivation, the first officer was primed to watch the captain closely. A vital part of good CRM is a thorough preflight brief to include the emotional, physical, and psychological state of each crew member. We need to express "where we are," especially if in a situation where we rarely fly with the same people. If you are fortunate to fly on a regular basis with the same crew members, you can often pick up from subtle cues where your flying mates are psychologically that day. Research suggests that crews who fly together more often are more confident as a team. One reason they are more confident is that they know each other better. Extending that idea, the more confident a crew is, the better it performs in the aircraft even when taking into account considerations like experience and ability (Smith 1999). However, most airline crews rarely fly with the same crew members other than on one trip. Had this captain chosen to hide his lack of sleep, the first officer may not have been so vigilant in his cross-check.

Speak up, son, I can't hear you

Why is it that crew members have notoriously had a difficult time speaking up to the captain? There are several possible explanations. The most obvious is that there exists a power differential between the captain and the other crew members. In the airlines, the captain can hold sway over your future employment at the airline, and employment helps to bring home the bacon. That is a pretty powerful reason to be careful of what you say and how you act. In the military, the aircraft commander (the equivalent of a captain in the airlines) can have a big impact on your military career, including the jobs you get in the squadron and ultimately where your next assignment will be and in what position. So clearly, there is a hesitancy in both the civilian and military crew member to make waves. The question becomes, when do waves become necessary?

Culturally, in the aviation business the captain of the ship has long held an aura of respect and dignity. "Who am I to question the captain?" was almost the mindset. On top of this, you may have been raised not to question authority. In this situation, you would be particularly susceptible to not speaking up in the cockpit. This idea of not questioning authority is becoming more uncommon in the United States today as children see posters in school that proclaim, "Question Authority!" Allow me to get on my soapbox for a moment. In my opinion the pendulum has swung a little too far in the United States as far as questioning authority goes. Authority is a good thing. Without authority, anarchy and mayhem reign supreme. Authority and obedience are a large part of what hold our system together. If everyone went his or her own way all of the time, we would have chaos. Chaos is the last thing we want in a cockpit. On the other hand, there is a healthy questioning of authority that we need to possess. When something seems amiss, we have an obligation to our other crew members and especially to our passengers to ask about it.

This concept of not questioning the actions of a superior is known as *professional deference*. We defer to the captain's professional judgment most of the time. However, like many things in life, deference runs along a continuum. We can get to the point of the continuum known as *excessive professional deference*. This point is where we become like sheep or yes-men or -women. We become very passive and do not engage in independent critical thinking. Or we are active but rely on the captain to do all our thinking for us. We need to become what Robert Kelly (1992) calls "exemplary followers" who are active and think critically. The Viking first officer cited in the preceding account was an exemplary follower, thinking and acting.

Folks who show excessive professional deference are hesitant to call out or question deficient performance or judgment on the part of the captain. If they do, it is done in a very vague way that may not be caught by the captain and/or in such a tone that even if caught by the captain, it may not register as a big deal to him or her. A good example of this is from UAL Flight 173 into Portland. The second officer made a vague reference to the critical level of fuel remaining by saying, "Fifteen minutes is really gonna run us low on fuel here." However, it seems that he took no affirmative action to ensure the captain was completely aware of how much time there was until fuel exhaustion. Vague references don't always get the job done. If they don't, they need to be followed with more direct comments. The accident board noted this and recommended that aircrew members get training in assertiveness. Crews are now taught to follow up subtle hints that go unheeded with statements such as this: "Pilot, can you explain to me why we are 10 knots slow?" If the pilot does not respond, the crew member is to ask again. Upon the second nonresponse, the crew member is trained to take the aircraft and execute the appropriate response.

Another classic example of vague (and not-so-vague) references that went unheeded is the Air Florida Boeing 737 that crashed into the icy waters of the Potomac River in 1982. The aircraft crashed because of excessive icing on the wings and engines. The copilot tried on several occasions to alert the captain concerning the abnormal engine indications they were receiving. The faulty indications were caused by the icing, and subsequently the crew set the throttles too low for a safe takeff, as the EPR (exhuast pressure ratio) indicated a false takeoff thrust setting. The first officer's alerts went unheeded by the captain, and subsequently the airplane crashed because the engines were not producing enough thrust (though they were capable of doing so.)

Milgram in the cockpit

Perhaps the most famous and powerful psychology experiment of all time was conducted by Dr. Stanley Milgram at Yale University in the early 1960s. The study has since been replicated with similar results in the United States and around the world. Milgram was Jewish, and he sought to understand how the German people could have allowed the Holocaust to occur. Were they simple morally inferior to the rest of the world, or do we all possess the capability to induce such atrocities? A disturbing thought. Milgram hypothesized that perhaps it was the power of the situation that drove people to such behavior. Conditions were perhaps set up that allowed this evil tendency to be displayed. Milgram enlisted a subject to teach a learner a set of word pairings. The subject was told that the researchers were interested in how we learn. If the learner (who was actually an accomplice of the researcher) failed to respond with the correct word pairing, the teacher was to administer a short electrical shock to the learner. Each time that the learner missed a word pairing, the teacher was to increase the voltage until he reached a level of above 400 volts, which was marked "Extremely dangerous, XXX." Before the experiment began, Milgram asked scientists to estimate how many subjects would actually administer high levels of shock. With a high consensus they estimated it would be only 1 to 2 percent of the people. The results were shocking. A full 70 percent of the people carried out their orders and administered the highest levels of shock available. The people did not do it joyfully but often looked quite pained and under duress as they increased the shocks. When they looked to the experimenter to stop the experiment, the experimenter simply replied, "The experiment requires that you continue." If pressed, the experimenter would say, "I take responsibility." With the use of these two simple lines, the majority of the subjects continued to administer the shocks. These experiments became known as Milgram's Obedience Study. It has been a much-scrutinized experiment for a number of reasons, not the least of which is its disturbing findings. I should add that the learner actually received no shocks, but for realistic effect, an audiotape was played of the learner screaming and asking for the experiment to stop.

Milgram's results have been applied to the cockpit. Eugen Tarnow labels it *destructive obedience* (Tarnow 1999). After reviewing recent NTSB accidents in the United States, Tarnow suggests that up to 25 percent of all plane crashes may be caused by destructive obedience. He goes on to draw four parallels between Milgram's obedience study and aircraft safety:

1. *Excessive obedience.* It seems clear that people are willing to inflict severe pain or cause serious injury in the name of following verbal orders. This translates to the cockpit, with a crew member willing to risk personal harm or harm to passengers and others in order to obey.

2. *Hesitant communications.* Subjects' objections to the experiment could easily be dissuaded by the experimenter simply saying "the experiment requires that you continue." Objections were usually made in a very hesitant tone and/or manner. In the aircraft we have already seen that hesitant or vague expression of concern is not always effective. If objections are raised in this tone, it may be easy for the captain (or others) to override or veto the objection unduly.

3. *Accepting the authority's definition of the situation.* Subjects readily accepted the experimenter's definition of the situation, which included the fact that obedience was required but disobedience was not even an option. Of the nearly 1000 trials, only one subject attempted to free the learner or call the police. While airborne, if we

are unsure of a situation, it may be easy to defer to someone who is willing to confidently paint his or her own picture of the situation.

4. *Closeness effect.* The experimenter's authority over the subject was much stronger the closer the experimenter was located physically to the subject. A pilot team sitting side-by-side is in very close proximity. It may be easier for someone further removed (e.g., the second officer) to offer a more objective and confident objection when necessary. Combating this strategy is the fact that most second officers are the most junior of the aircrew and the most vulnerable to having their career hurt by a captain. In military aircraft, sometimes crew members are separated by compartment (e.g., F-14 pilot and WSO, B-1 offensive and defensive operators from the pilots). Use this physical separation to your advantage.

Check the Winning Number

The flight was a MD-80 en route to Anchorage (ANC). Jim was the captain and Tony the first officer. During the initial portion of the en route descent, Tony warned Jim about a low altimeter setting in ANC. It was cold and IMC in Alaska's largest city that January afternoon. As the crew descended through FL180, the crew set its altimeters and did its cross-check and approach check. The captain, Jim, set his altimeter to 29.69 (incorrect), and Tony set his altimeter to 28.69 (correct). At that cross check with 29.69 set, 17,800 ft would have been displayed on Jim's altimeter. At that same time, Tony's would have 28.69 set, with 16,800 ft displayed. The first officer (PNF) should have caught the discrepancy but didn't. Jim had dialed in the wrong setting. No one mentioned a thing. As the crew continued down, it was cleared to 10,000 ft. While descending through 10,000 ft on Tony's altimeter, Tony looked over to crosscheck with Jim's altimeter, which displayed about 10,600 ft. Tony made a verbal utterance while trying to determine whose altimeter was correct. Jim realized Tony was indicating something was incorrect and immediately disengaged the autopilot and initiated a climb to 10,000 ft on Tony's altimeter, bottoming out at 9600 ft MSL. Right then, ATC asked for the altitude, and the crew reported it was climbing back to 10,000 ft. The minimum en route altitude (MEA) was 9000 ft and the minimum obstruction clearance altitude (MOCA) was 8500 ft. (ASRS 425460)

The little things

Jim and Tony present a classic case of overlooking a minor detail that has major ramifications. Tony had warned Jim of a low altimeter setting in Anchorage. Jim indeed set 29.62, which is lower than the standard 29.92. They had a standard and widespread procedure of setting in the new altimeter as they approached FL180. They even had a procedure to cross-check altimeter readings after the settings were put in, looking for discrepancies in altimeters. Usually a discrepancy of about 100 ft will begin to get a pilot's attention. In this case, they had a 1000-ft difference, but neither pilot caught the discrepancy during the cross-check. Both pilots share the blame. Jim actually dialed in the wrong setting, while Tony was in a better position to catch the error on the cross-check as he was the pilot not flying (PNF). The PNF is the person to back up the pilot flying who is more in the heat of the battle.

Procedures and checklists have no value if not followed. This is simply a case of a lack of attention to detail. In military flying, attention to detail is preached ceaselessly and with good reason. The actions of Jim and Tony are excellent examples of how a small detail (one digit off in an altimeter setting) can have large consequences on performance (a 1000-ft altimeter readout difference). That is one attribute of flying that makes it such an interesting business.

Usually, large differences in readout are more easily noticed by the aircrew, but not this time. In fact, the large size of this difference was actually less likely to be caught in this situation. The reason is simple; it just so happened that a 9 instead of an 8 would cause a 1000-ft difference, while a 700-ft difference would have more easily caught the first officer's eye. A altitude of 17,800 would look very similar to an altitude of 16,800 on both a digital and an old, round dial altimeter. In the case of the digital read-out, only one of the middle digit differs. Digits at the beginning and end tend to catch the eye better than digits in the middle. We as humans are also trained to look for patterns, and the pattern of the two numbers would appear very similar. The same goes for the old round dials. The "little hand" would have appeared stable in a very similar position (6 versus 7 on the clock), while the "big hand" would have been in identical locations for both 17,800 and 16,800. Finally, depending on the angle of viewing (parallax), lighting conditions, and eye fatigue, it could have been difficult for the PNF to detect the difference, especially with a quick scan, as is often the case on a routine checklist item. Visual acuity could have been difficult. It's a simple mistake all of us have or could have made. The lesson is this: Little errors can have big consequences. The remedy, of course, is vigilance and attention to detail on all checklists.

Groupthink

Another reason that crew members may be unwilling to voice questions concerning the validity of a decision is a concept known as *groupthink*. Groupthink is one of those psychological terms that has slipped into the American mainstream vernacular. The term was first introduced by Irving Janis in 1972 to explained major decision-making fiascos such as John Kennedy's decision to go ahead with the Bay of Pigs invasion, but to call off air support at the last minute, costing hundreds of the lives of those arriving by sea. Janis coined his term from George Orwell's famous book *1984*. What is interesting is that the validity of groupthink has recently come under attack by some in the scholarly community. While their arguments are beyond the scope of this book [see Fuller (1998) for more information], it is important to point out that group consensus does not always lead to bad decisions. In fact, many times it may lead to good decisions.

Several definitions of groupthink have been offered since 1972. For the purposes of our discussion, I will define groupthink as an overcommitment to consensus by individual members of a group. This desire to not rock the boat can lead us to bury our uneasy feelings and concerns about where the group is going with certain decisions. Often we may not be concerned with the actual decision, but, rather, we feel uncomfortable about how the process is being conducted. Gregory Moorhead and colleagues (1994) offer several reasons why groupthink can occur and several ways to avoid it. The prerequisites for groupthink are having a very cohesive group, where the leader makes his or her preferences clearly known, and when the group is insulated from outside experts. The group begins to feel invulnerable; it sterotypes the views of others, puts pressure on dissent, self-censors, and has illusions of unanimity. The key to combating groupthink is for the leader to be strong and demanding by forcing critical appraisals of alternatives. The leader should take on the role of devil's advocate and ensure that others do too. The leader must shield the group as much as possible from the pressure of time. Try to extend the time available for a decision if possible. If not, force the group's attention to consider the issues rather than time and remember to withhold your own personal preferences until the group has shared how it feels. These make good recommendations for a captain or aircraft commander.

Remedy and summary

The remedy to many CRM problems is simply to speak up and voice your concern. Worry about the consequences of speaking up later on. Most professionals will appreciate that you spoke up. True professionals are secure enough to realize that they make mistakes and your question is not twisted into a question of their abilities. If they aren't professional and don't appreciate your comments, you can't be overly concerned about that. That is their problem, not yours.

Social loafing

Besides groupthink, another common reason people fail to speak up, or even act, is known as *social loafing*. If you ever had a group project to do at school or work, you are probably familiar with this concept. Social loafing is basically where certain members of the group are content to sit back and let others (often a few) do all of the work. Social loafing is group members ducking responsibility and leaving the work to others. Those who really care about the group producing are willing to put in the work, while others are happy to ride their coattails.

Studies have shown that social loafing is most likely to occur as group size increases. When team members feel lost in the crowd, they don't feel accountable for their actions and aren't concerned about being evaluated. We can see this in the cockpit by contrasting an aircraft with six crew members, like the B-52, with an aircraft with just two crew members, like the B-2. In the B-2, it will become obvious very quickly that the one other person on board is not pulling his or her weight. The individual B-52 crew member can hide behind the coattails of others for a while.

There are some very effective ways to combat social loafing, including getting crew members to be more aware of how they are doing, perhaps by comparing their performance with a standard. Having others evaluate individual performance is also a way to increase effort. Providing group rewards, encouraging group commitment and the belief that one's teammates are working hard, and having the group engage in a challenging task are all ways to combat social loafing. The passenger syndrome discussed in Chap. 4 is a type of social loafing; though the actor may be more physically active, mentally he or she is relying on the authority figure to bail him or her out.

I would hope with a group of professional flyers that social loafing would be the exception rather than the rule. Social loafing is a sign of immaturity and is hopefully rare in the flying community. Having said that, we all have a tendency to be lazy at times. So we could all be susceptible to this phenomenon. While it should not be the pattern of our flying life, we must not let it creep in even on an infrequent basis, though we may be tempted to do so.

Peanuts and a Coke

Many problems with CRM start with the captain. If his or her leadership is strong, then CRM has a chance to be really good. If leadership is poor, CRM has an uphill battle to fight. Part of effective leadership is asking the right questions. This is clearly demonstrated by our next case study (adapted from the book *Unfriendly Skies*).

The crew was flying into Saginaw, Michigan. The captain was 42 years old, but most of his flying had been in the southern portion of the United States, so he had never flown into Saginaw. He had been with this airline his entire career. Like so many times in this day and age, his company had recently absorbed a smaller airline. When he met his fellow crew members that day, he found that they had been employees of the now-absorbed airline. The flight engineer appeared young and eager. The first officer was a bit older—in fact, the same age as the captain.

The crew departed Miami on the first leg of the trip and the captain did the flying. As per custom, the first officer had the honors of flying the second leg. The captain was not terribly comfortable on the 727, as most of his experience in this equipment had been of a training nature. So the captain asked the first officer how long he had been flying the Boeing 727. The first officer replied, "Eight years," to the captain's satisfaction. He followed up with another question on whether the guy had been to Saginaw before. "Been there since the day I joined the company" was the first officer's slightly smug reply.

Saginaw was undergoing some airport renovation. The longest runway had been shortened from 6800 ft to 5500 ft. The secondary runway, 18, was 5000 ft long. A 7000-ft runway was considered average for the 727. Getting down to around 5000 ft made things pretty tight (though within limits) for the bird. At that short of a distance experience and confidence in the aircraft was vital to operating the jet by the book in order to stop the aircraft within the confines of the runway. On this day, they were calling winds 180 at 25 knots, a significant headwind dictating that the shorter runway be used. The bonus of course was that a 25-knot headwind would significantly slow ground speed, easing the problem of stopping within the runway distance.

The captain queried the copilot as to how he planned to handle the landing. The captain is ultimately responsible for the safe conduct of the flight and wanted to ensure his first officer had a good game plan. The first officer planned a 40-degree approach. This would allow an approach speed of roughly 120 knots, which would further aid in landing on a short runway. Forty flaps would also result in a lower deck angle (lower flaps = lower deck angle). The standard flap setting in the southern states was 30 flaps, but that would require a faster approach speed (because of less lift produced by the wing), which would increase stopping distance. The captain was very happy with the plan. He was pleased to have such a skilled first officer with him on this flight.

The flight was about 100 s from landing. The captain sat blissfully with a bag of peanuts and a Coke at his side—it was clear sailing. As the captain looked out the front window, he was privileged to have a view he had seen thousands of times out the nose of the aircraft toward terra firma. It is something that is exhilarating no matter how many times you have seen it: coming toward a patch of concrete in a big hunk of metal with a load of passengers behind you. Though it is exhilarating, you get used to the sights and the sounds. The sound of the engine, the look of the instruments, the picture outside all become well ingrained. But this time it didn't look so familiar.

The captain began to ponder how things looked so different at this lower deck angle (caused by the change from 30 to 40 flaps). He began to feel uncomfortable, buoyed only by the fact that he had an experienced first officer at his side. He rolled a peanut on his tongue. He had this sinking sensation that the plane was going to crash, but that couldn't be the case.

The trees seemed to be growing larger and larger; it didn't seem like the plane would clear them. If the trees were cleared, it seemed like the lights would be insurmountable. The level of uncertainty (known as the *pucker factor*) was rising in the captain, contained only by the knowledge that the first officer had things under control, that he was experienced.

Finally, an unprintable utterance sprang from the captain's lips as he grabbed the yoke, yanked it back, and added full power. The plane was landing several hundred feet short of the runway! He yelled for full power; the only chance to clear the lights was to "jump" over them. The throttles were plastered against the firewall. The plane jerked mightily. Crashes of trays were heard in the back of the aircraft. The airspeed jumped from 122 to 143 knots, and the aircraft lumbered onto the tarmac. Later analysis revealed the aircraft cleared the end of the runway by 30 inches (not a misprint)!

The captain then yanked the throttle to idle and deployed the spoilers and thrust reversers. He then jumped onto the brakes with all his might. The aircraft went screaming down the runway, skipping and skidding. Overhead bin contents tumbled, coats rained from the closet, oxygen masks dangled in front of passengers, the plane rolled on, and the end of the runway drew near. Suddenly there was silence. The aircraft came to rest with the nose peeking over the edge of the runway. Somehow the captain managed to get the aircraft turned around and taxied to the gate: "Thanks for flying with us today." After the passengers deplaned, there was a dead silence in the cockpit.

After two to three minutes, the captain suggested that the second officer go get some coffee. The flight engineer gladly agreed and promptly obeyed. The conversation that unfolded next could be described as frank, straightforward, heated, and mostly one-way. The first officer received a stern and well-deserved lecture. During the short Q and A session, a small feature of this conversation, the captain learned that the first officer had not told the captain one thing that was false, but he was misleading. He did have eight years in the 727 and he had been to Saginaw many times—but only as a flight engineer. This was only his third trip as a first officer, and he had never accomplished a 40-degree flap landing in the aircraft!

Booby traps

The captain went on to point out that had this aircraft crashed (a very real possibility), the accident investigation would have undoubtedly labeled the crash as pilot error. But pilot error is often simply a booby trap that we fall into when we don't have to. There are several lessons to be drawn from this case. The captain did a nice job of enumerating many of them.

Both the flight engineer and first officer were new to the captain. This is not an entirely unique situation, especially within a large airline. Often crew members are unfamiliar with one another. However, this adverse situation was compounded by the fact that both these crew members were also new to the company, having come from a smaller and perhaps less professional airline. They would not be as familiar with company aircraft, procedures, and protocol. They would also be eager to prove themselves worthy to the new company. Like many in a new job, they would be very unwilling to reveal any chinks in their armor.

Hidden agendas

Some have called this a problem of hidden agendas. Hidden agendas can lead to crew decisions that satisfy certain individuals on the flight. A classic is get-home-itis, where one crew member may desperately want to get home so he or she manipulates other crew members into flying an approach they really shouldn't fly. The Saginaw copilot seemed to want to protect his ego at the expense of the crew. In this case, the copilot's hidden agenda—to appear to be a competent and good employee—backfired. He ended up looking very silly. Exposed hidden agendas almost always result in shame.

Coke and peanuts syndrome

The copilot didn't tell the captain anything that was not true. However, he retained certain information that would have been very helpful for the captain (and second officer) to know. Had the captain been aware of the actual state of the copilot's experience, he would have been more apt to discuss techniques that may have helped the crew avoid the disastrous approach. The captain would also have been much more likely to avoid what I call the Coke and peanuts syndrome, where we sit there fat, dumb, and happy because we have an experienced crew member at our side with everything under control.

The Coke and peanuts syndrome is a type of social loafing, discussed earlier. Therefore, Coke and peanuts is closely related to the passenger syndrome, which we discussed in Chap. 4. The passenger syndrome is the opposite of Coke and peanuts in that the person not in authority is relying on the authority figure to bail him or her out (i.e., "I'm just along for the ride; whatever happens, he or she will bail me out"). The Coke and peanuts syndrome, on the other hand, involves an authority figure relying on those below him or her with a misplaced overconfidence. Granted, we cannot micromanage our crew; there must be a level of trust that the other crew member will do his or her job. However, we should all possess a healthy cross-check of each other.

As stated earlier, it is the captain's responsibility to ask the questions necessary to ascertain the true state of his or her crew. It is the crew member's responsibility to volunteer the information the captain needs to make that assessment. Both the passenger syndrome and the Coke and peanuts syndrome are subcategories of social loafing.

Ask good questions

At the same time, the captain bears responsibility as well. As he states himself, "I realize how silly it was. I realize how foolish I was to have placed so much confidence in a man I didn't know and whose only collateral was his self-assured arrogance." The captain exhibited a failure of good leadership. A good leader completely assesses a situation. The captain asked good questions of the copilot, but not enough of them. He didn't ask the number and kinds of questions required to ascertain an accurate picture of his copilot. Questions such as "tell me about your flying career with Company X," "tell me what you have done since you joined our company," and "what kind of training and experience have you had?" would have shed valuable light on the nature of the copilot's flying abilities. With an accurate assessment of the copilot, the captain would have much more likely intervened earlier in the botched approach. Perhaps he could even have talked the copilot through the approach. This was not possible under the Coke and peanuts syndrome.

Another reason that the captain failed to query the copilot in detail is that they were both of the same age, 42. It is human nature to treat someone of the same age (or nearly the same age) as a peer. The hierarchical structure of an aircrew is often made smoother when the captain is clearly older than the first officer, who is older than the second officer, and so on. This is not always the case, however. Many times the airline may have hired a retired military flyer who will "sit sideways" and take orders from a captain and first officer who are much younger. Keep in mind that it is much easier to ask questions with an air of legitimacy if one crew member is clearly older than the other. This is just human nature. It is part of a natural power differential. When the age and power differential are out of balance, it can strain the hierarchical system (but it doesn't have to). Ask the right questions and don't make assumptions is a good motto for any captain.

Unfamiliarity

The captain was unfamiliar with the airport. This event occurred right after the deregulation of the airlines ordered by President Jimmy Carter. Prior to deregulation, pilots often flew the same routes on a routine basis. Deregulation changed everything. Routes and destinations became competitive, and airlines were working hard to get as many routes as possible. As a result, pilots went to new places routinely. This, of course, would not be as big a problem in today's environment. However, it is not unknown. Pilots and captains do travel to destinations for the first time. Today we have better intelligence on airports, with technology in the cockpit displaying up-to-date airport information. The Internet also provides us with a host of information on our destination. Even with these advantages, as the old saying goes, nothing beats having been there. The fact that the captain was unfamiliar with the airport eased his transition into the Coke and peanuts syndrome. The copilot had been there, and that was invaluable in the captain's eyes.

One final word on our friends at Saginaw. Clay Foushee and colleagues (1986) conducted a study where they compared the simulator performance of crews that had flown together for several days (Post-Duty) to a fresh crew just reporting for duty as strangers (Pre-Duty). "The results revealed that, not surprisingly, Post-Duty crews were significantly more fatigued than Pre-Duty crews. However, a somewhat counter-intuitive pattern of results emerged on the crew performance measures. In general, the performance of Post-Duty crews was significantly better than that of Pre-Duty crews, as rated by an expert observer on a number of dimensions relevant to flight safety. Analyses of the flight crew communication patterns revealed that Post-Duty crews communicated significantly more overall, suggesting, as has previous research, that communication is a good predictor of overall crew performance. Further analyses suggested that the primary cause of this pattern of results is the fact that crew members usually have more operating experience together at the end of a trip, and that this recent operating experience serves to facilitate crew coordination, which can be an effective countermeasure to the fatigue present at or near the end of a duty cycle." It seems clear from this study that crews are most vulnerable to poor crew coordination the first day of a trip, even though they are more rested.

That is not a coincidence. A crew's unfamiliarity with one another and resultant lack of awareness of each other's abilities is likely a factor in the higher accident rates we see on the first day of a crew's trip. Additionally, research has shown that in early stages of a team's development, members are more likely to keep conversation at a congenial, polite level and are very adverse to engaging in argument. Therefore, to risk bringing up a challenge of another crew member, let alone the captain, is very unlikely. The scenarios we have been discussing are all brought together in another classic case study of poor CRM—Hibbing, Minnesota.

Hibbing, Minnesota

It was December 1, 1993. Dennis was the captain of Northwest Airlink Flight 5719 en route to Hibbing, Minnesota. His first officer was Kent, age 25. The two had flown trips together on two separate occasions in the past three months. Their day had begun at 1325 central standard time (CST). They met at Minneapolis/St. Paul International Airport (MSP). Both pilots were on reserve, so they deadheaded aboard the aircraft to International Falls, Minnesota (INL). From there they took over the same aircraft, a Jetstream BA-3100, and flew it back to MSP. Once the aircraft was turned and loaded, the captain flew the next segment of the flight. The aircraft was heading back to INL, with an intermediate stop at Hibbing, Minnesota (HIB).

The weather in Hibbing was forecasted to be 800 overcast with a visibility of 3 mi with light freezing drizzle and wind 180 degrees at 12 knots upon the arrival at Hibbing. As the plane approached Hibbing, the weather observation was ceiling of 400 overcast, visibility of 1 mi, light freezing drizzle, light snow, fog, and a temperature of 29°F. Winds were 180 degrees at 10 knots.

Approach initially wanted to give the aircraft the ILS to Runway 31. The crew talked it over and decided to land on Runway 13 because of the tailwind on Runway 31. So the crew requested the back course localizer approach to Runway 13. The controller cleared it for that published approach, which the crew initiated by joining the 20 DME arc. As the captain, Dennis, was completing the arc, he was at 8000 ft. Once the crew intercepted the localizer, Dennis briefed that he would begin descent. However, when he actually intercepted the localizer course, he asked the first officer what altitude he could descend to. The first officer replied 3500 ft. However, the captain remained at 8000 ft. After 9 s the first officer asked him if he was going to remain at 8000 ft. The captain said yes, presumably to avoid icing conditions. They were approximately 19 NM from the airfield. They had roughly 6650 ft to descend, as the field elevation is 1353 ft. The crew began to configure the aircraft, with gear and flaps to 10 degrees, and began to get ready for the before landing check. At approximately 17.5 DME, the captain began a descent. The average descent rate was 2250 ft per minute. The first officer asked the captain if he wanted him to run the before-landing check. The captain started to call for the before-landing check, but changed his mind and told Kent to run it once they crossed the final approach fix (FAF). When the aircraft passed over the FAF, it was 1200 ft above the minimum altitude.

At 3000 ft the first officer called "one to go" until they hit the step-down altitude of 2040. The first officer then called "twenty forty to, ah, ten point oh." The captain then asked the first officer if he had clicked on the airport lights. The plane descended through the step-down altitude at 2500 ft per minute. About 10 s later, the captain again asked about clicking the lights on. The first officer said, "Yup, yeah, I got it now." One-half second later, scrapinglike sounds were heard as the aircraft hit the top of the trees. The scrapping sound lasted for 3 s. (NTSB B94-910407)

Rapid descent

The accident board found several factors that led up to this CFIT. The first was the unstable approach flown by the captain. The captain elected to remain at 8000 ft on the approach when he was cleared to 3500 ft. By delaying his descent, he was forced to make a very rapid descent on the localizer approach. Company guidance suggested a descent rate of at least 1000 ft per minute on a nonprecision approach. However, this captain averaged a descent rate of 2250 ft per minute. Even at this excessive descent rate, he crossed the FAF fix 1200 ft above the minimum altitude, forcing him to continue the high rate of descent in order to reach the minimum descent altitude (MDA) before the visual descent point (VDP). Interviews with company pilots revealed that it was a standard practice for the pilots to accomplish rapid descents through icing conditions. However, the BA-3100 is certified for continued operation into known icing conditions. The procedure violated the manufacturer's and company's guidance. It was also a violation of the concept of flying a stabilized approach.

Organizational factors

As James Reason points out, there are many lines of defense in aircraft mishap prevention. The last line of defense is the pilot. However, the organization can set up the pilots for failure by either poor procedural guidance or a failure to enforce standards. This seems to be the case with the unauthorized practice of rapid descents in an aircraft certified to fly in icing conditions. The company had a culture where this could occur.

The company also failed to provide supervision to a captain who had repeatedly demonstrated poor performance. The captain had failed three proficiency check flights. In every one of those checks, his judgment was listed as unsatisfactory. Two of the checks had noted that his crew coordination was unsatisfactory. The failures occurred 6 months, 15 months, and 5 years prior to the accident. In 1989 the captain was subject to a 3-day suspension for negative reports relating to his flight performance. In addition to these documented flight deficiencies are several interpersonal problems noted in the following. Despite this performance, the company continued to let the pilot fly with no increased supervision.

Lifestyle stress

Lifestyle stress is stress that is produced outside of the cockpit. It can be long-term and chronic and can be caused by positive factors (e.g., marriage) or negative factors (e.g., high debt). The captain seemed to be experiencing large levels of stress with the company. In fact, he told several fellow pilots that he was angry with the company. Recently, the company had decided to base some of its pilots at some outlying areas where the company flew. Because of this decision and even though he had been with the company for six years, he was forced to give up his captain's position on a larger aircraft, the SF-340, and begin flying as a reserve captain on the BA-3100. This resulted in a 12 percent cut in pay, though he was able to stay in his domicile of preference, Minneapolis. In the past year and a half, the captain had filed three grievances against the company through the Air Line Pilots Associa-tion. He felt he had been a target for attention by company management. When he showed up in the pilot's lounge the day of the flight, another pilot said that the pilot was unhappy and had bellowed, "They violated my contract again!" He felt the company was forcing him to fly on a scheduled off day. Lifestyle stress can be carried over into the cockpit, making normal flight stress more acute. It can create an atmosphere of hostility from the moment the crew assembles.

On the day of the flight, the captain had a run-in with a customer service agent. He blew up when she asked him to make a call to confirm his authorization to dead-head on the first flight (this was a standard procedure). The captain called the chief pilot to complain, and the customer service agent's supervisor insisted that she file a formal complaint against the captain. He had also yelled at her in the office one month prior.

Not a nice man

As mentioned previously, the captain had twice been cited for poor crew coordination during check flights. One evaluator talked to the captain about the problem. He said that for the remainder of the simulator training days the captain improved, but he felt it may have been a "cooperate and graduate" move on the part of the captain. It seems likely that this was the case. Many described the captain as above average in intelligence with a sharp wit. Five of the six pilots who had flown with him as his first officer reported being intimidated by him. One first officer reported being struck by the captain when he inadvertently left the intercom on. During the day of the flight, the captain chewed out the first officer, in front of a ramp service agent, for incorrectly checking the exterior lights. The ramp agent relayed that there was anger in the captain's voice and that the first officer appeared embarrassed. Additionally, there were four formal complaints of sexual harassment involving the captain during 1988 and 1989. A passenger filed a formal complaint when he noticed the captain sleeping in the cockpit during a flight. He had also been cited by the company for flying with mechanical irregularities.

The accident investigation board determined that the probable cause of the accident was the captain's actions, which led to a breakdown in crew coordination and the loss of situational awareness by the flight crew during an unstabilized approach in night instrument conditions. The captain's actions and questions led to distraction during critical phases of flight. They distracted the first officer from his duties of monitoring the captain's altitude and approach. The captain's timing and instructions illustrated the worst in crew coordination.

Robert Ginnett studied the interaction of aircrews as they flew three- and four-day trips together. He determined that the tone of the mission can be set in as little as 10 min by the captain's action and attitudes. The pilot in the preceding account is a classic example of a captain who continually set the wrong tone and it caught up with him and his crew.

Overbearing

While the essence of the personality of the captain of UAL Flight 173 into Portland remains unclear, the essence of the Hibbing, Minnesota, captain does not. Clearly, he was an overbearing ogre. It is difficult enough to bring up dissent in a neutral cockpit; the social pressure in a hostile aircraft makes it almost unthinkable. To put it bluntly, an overbearing commander can create an environment ripe for disaster aboard any aircraft (read: it can lead to death). Sometimes this is the cause of poor CRM, pure and simple. We discussed excessive professional deference earlier; an ogre makes this much more likely to occur. We looked at a form of poor leadership in the Saginaw case. That was poor leadership because of a lack of proper questioning. That lack of questioning was the result of an intent of being a supportive commander. The type of questioning and environment at Hibbing was with the intent of intimidation. Both are a failure of leadership. I will let you decide which is the more dangerous approach and urge you not to venture to either style of leadership.

Ginnett (1995) has studied the formation and effectiveness of airline crews for a number of years. He has observed that highly effective captains (and other leaders) often use a variety of authority techniques in the first few minutes of the crew's life. These intial moments are key in setting the tone for the entire trip.

Bob Helmreich at the University of Texas has done the lion's share of the work in the field of CRM. Helmreich states that the communication value that most clearly separates excellent copilots and poor copilots is a willingness to question the decision or actions of the captain, clearly a value of openness. An air of intimidation makes it difficult to encourage the copilot to question decisions. Why is this so? Karl Weick (1990), another CRM researcher, adds that stress changes communication from horizontal to vertical. In other words, it changes from a dialogue among professionals to a direct, one-way communication order from the captain. In some situations this is appropriate. Certain emergency situations (e.g., a fire light on #2) require immediate and well-trained responses, not a conversation. However, most emergencies require good crew coordination, rather than immediate response. Or if the initial state of the aircraft requires a quick response, once the response is made, there is usually time for a thorough discussion of the next move. The UAL DC-10 Sioux City crew mentioned at the beginning of this chapter was able to combat a totally unforeseen emergency rather capably (actually, remarkably well) primarily because of horizontal communication and teamwork. For example, the captain had the flight attendant invite another UAL pilot sitting in the back forward to bring him into the conversation. In the end, he proved invaluable. None of this occurs with a tyrant for a captain. Weick goes on to point out that horizontal communication is needed to correct errors and to detect false hypotheses by speaking out loud. He contends that 84 percent of accidents were caused by a failure to monitor or challenge a faulty action or inaction.

References and For Further Study

Foushee, H. C., J. Lauber, M. Baetge, and D. Acomb. 1986. Crew factors in flight operations. III. The operational significance of exposure to short haul air transport operations. NASA Tech Memorandum 88322.

Fuller, S. R. and R. Aldag. 1998. Organizational tonypandy: Lessons from a quarter century of the groupthink phenomenon. *Organizational Behavior & Human Decision Processes,* 73:163–185.

Ginnett, R. C. 1987. First encounters of the close kind: The formation process of airline flight crews. Unpublished doctoral dissertation, Yale University, New Haven, Conn.

Ginnett, R. 1993. Crews as groups: Their formation and their leadership. In *Cockpit Resource Management.* Eds. E. L. Wiener, B. G. Kanki, and R. L. Helmreich. Orlando, Fla.: Academic Press.

Ginnett, R. C. 1995. Groups and leadership. In *Cockpit Resource Management.* Eds. E. L. Wiener, B. G. Kanki, and R. L. Helmreich. San Diego: Academic Press.

Janis, I. L. 1972. *Victims of Groupthink.* Boston: Houghton-Mifflin.

Kelley, R. 1992. *The Power of Followership.* New York: Doubleday Currency.

Kern, T. 1997. *Redefining Airmanship.* New York: McGraw-Hill. Chap. 6.

Milgram, S. 1974. *Obedience to Authority: An Experimental View.* New York: Harper & Row.

Miller, D. T. and C. McFarland. 1987. Pluralistic ignorance: When similarity is interpreted as dissimilarity. *Journal of Personality and Social Psychology,* 53:298–305

Moorhead, G., R. Ference, and C. P. Neck. 1991. Group decision fiascoes continue: Space Shuttle Challenger and a revised groupthink framework. *Human Relations,* 539–550.

Reason, J. 1990. *Human Error.* Cambridge, England: Cambridge University Press.

Smith, D. R. 1999. The effect of transactive memory and collective efficacy on aircrew performance. Unpublished doctoral dissertation, University of Washington, Seattle, Wash.

Tarnow, E. 1999. Self-destructive obedience in the airplane cockpit and the concept of obedience optimization. In *Obedience to Authority: Current Perspectives on the Milgram Paradigm.* Ed. T. Blass. Mahwah, N.J.: Lawrence Erlbaum & Associates. pp. 111–123.

Tuckman, B. W. 1964. Developmental sequence in small groups. *Psychological Bulletin,* 63:384–399.

Weick, K. E. 1990. The vulnerable system: An analysis of the Tenerife air disaster. *Journal of Management,* 16:571–593.

Wortman, C. B. and E. F. Loftus. 1992. *Psychology,* 4th ed. New York: McGraw-Hill. Chap. 19.

X, Captain and R. Dodson. 1989. *Unfriendly Skies: Revelations of a Deregulated Airline Pilot.* New York: Doubleday.

Zimbardo, P. G. 1974. On "obedience to authority." *American Psychologist,* 29:566–567.

6

Automation

Automation is supposed to make flying easier. That's not always the case, but automation has certainly been a benefit to the flyer in many ways. Automation has enhanced flight safety. It has also increased the efficiency of flight operations. It has helped reduce the workload on the pilot and cut down on the resource demands of certain tasks (e.g., flying a fix-to-fix to a holding pattern). Indeed, it would now be impossible to fly certain aircraft without the aid of automation. The F-16 Fighting Falcon and B-2 Stealth Bomber are prime examples.

Chris Wickens at the University of Illinois has written extensively on the subject of aircraft automation. He points out that automation can take one of three different forms:

1. Automation can collect and represent information for the pilot.

2. Automation can choose or recommend actions for the pilot.

3. Automation can also execute those actions.

Essentially, automation can play two different roles. First, automation can augment the pilot (e.g., provide automated checklists). The philosophy behind this approach is that automation is to work in parallel with the pilot and allow the human to use the information as desired. Second, automation can replace pilot information processing (e.g., compute optimum routing). The underlying philosophy in this case is that automation will do the task in place of the pilot. Pilots are very wary of this second role or philosophy. As one pilot put it, "I am willing to fly it as long as it has the yellow button (autopilot disengage). I can always turn it back into a DC-9" (Wiener 1985, p. 88). This idea rests in a psychological concept known as *perceived control*. In general, people feel less stressed if they feel like they have control in a situation, even if in fact this control is lacking. That is why the dentist tells you, "If this root canal hurts too badly, just raise your hand and I will stop what I am doing." Pilots feel the same way about automation; they are willing to lay there and get the benefits, but they want to be able to raise their hand and stop it if it gets too painful. Pilots don't like feeling that they can be replaced by a machine.

Three Big Inventions

There have been three giant strides in aviation automation over the last 30 years. (Some would consider the low-level wind shear device, discussed in Chap. 2, a fourth great stride.) The first is what is known as the Ground Proximity Warning System (GPWS). GPWS is now in its fourth generation. The GPWS is designed to protect against terrain collision. The GPWS has employed a variety of inputs in its history. Through radar and/or radio altimeters and/or onboard mappings through the Inertial Navigation Systems (INS) and the Global Positioning

System (GPS), the GPWS computes an aircraft trajectory to determine if the aircraft is in danger of ground impact. If the system determines that such an impact is imminent, a "pull up, pull up" (or similar) audible warning is sounded to alert the aircrew.

The second product is known as an FMS (Flight Management System) or an FMC (flight management computer). The FMS is a complex system of autopilots that can provide automated action for both aviation and navigation. The FMS can compute optimal trajectories through the airspace and actually fly the aircraft to this point at a certain time and altitude. With such a device, pilots are able to calculate precise arrival times using current winds, temperature, and so on. Furthermore, the system gives an accurate picture of fuel burn rates and the center of gravity (cg) for the aircraft. In essence, the FMS is the brains or nervous system of the entire modern jet airliner, business jet, or military aircraft.

TCAS is the third revolutionary automation product to change the way we fly. TCAS stands for Traffic Alert and Collision Avoidance System and is similar in concept to GPWS. However, instead of ground collision, TCAS is concerned with midair collision with other aircraft. Unfortunately, one of the accidents that precipitated the introduction of TCAS occurred in 1987, when a Pacific Southwest Airlines 727 collided in midair with a Cessna 172 on approach to Lindbergh Field in San Diego while under radar control. TCAS, through the use of radar, calculates a trajectory through three-dimensional space. In fact, it calculates the position and future trajectories of various aircraft in close quarters. There is a display located in the cockpit portraying the immediate area and highlighting potential conflicts. If minimum separation is thought to exist, it offers the pilot a suggested maneuver to avoid the conflict. TCAS is an excellent example of an automated device that augments the pilot (in this case his or her eyes) in a necessary flight function: clearing for other aircraft.

Armed with this information on the purpose of automation and understanding some important recent innovations, we turn our attention to understanding how automation is related to CFIT and CFTT. Thus far, we have examined some classic aircraft accidents; the Portland, Oregon, and Hibbing, Minnesota, cases provide a wealth of lessons to those who still grace the skies. Perhaps the most famous aircraft accident of all time is Eastern Airlines Flight 401 in December 1972. It was an accident caused in part by a problem with automation. It is to this case we now turn our attention.

How Are Things Comin' Along?

It was December 29, 1972. It had been an eventful leap year. The Winter Olympics were held in Sapporo, Japan. The Summer Games in Munich featured Mark Spitz, Olga Korbut, Jim Ryun, and, unfortunately, a group of terrorists holding several of the Israeli athletes hostage. Richard Nixon had won reelection over George McGovern in a landslide the previous month. The Christmas bombings were under way over Hanoi in an effort to bring an end to the Vietnam War. Meanwhile, in New York, 163 passengers shuffled aboard an Eastern Airlines Lockheed L-1011 bound for Miami (MIA). Many of the passengers loading onto the aircraft at John F. Kennedy International Airport (JFK) were undoubtedly returning from Christmas holiday travels. Eastern Flight 401 departed JFK at 2120 EST that night.

The flight proceeded uneventfully until the approach to Miami. The captain ordered the landing gear to the down position. The crew waited for the green lights to indicate the gear was down and locked. One green light, a second, but the nosegear light failed to illuminate. The crew recycled the gear with the same result: no green light for the nosegear indicating down and locked.

At 2334:05 Eastern 401 called Miami Tower and advised it that the flight needed to go around because of no light on the nosegear. The tower acknowledged the call and had Eastern 401 climb straight ahead to 2000 MSL and contact Approach Control. Once Eastern 401 contacted Approach Control, the flight relayed its status of no nosegear indication. Approach Control affirmed the call and gave the aircraft a vector to heading 360 and told it to maintain 2000 ft MSL. One minute later, the captain instructed the first officer to engage the autopilot. The first officer was flying the aircraft at the time, acknowledged the instruction, and engaged the autopilot.

Shortly thereafter, Miami Approach Control gave the aircraft a heading of 330. Meanwhile, the first officer successfully removed the nosegear light lens assembly, but when he attempted to reinstall it, it jammed. At 2337:08 (3 min after the first call to Miami for the go-around), the captain told the second officer to go down into the electronics bay located below the flight deck and visually inspect the alignment of the nosegear indices. The physical alignment of two rods on the landing gear linkage would indicate proper nosegear extension. If the nosewheel light is illuminated, the rods can be viewed through an optical sight located in the forward electronics bay. Before the second officer went down into the bay, the aircraft experienced a momentary downward acceleration of 0.04 g, causing the aircraft to descend 100 ft. The descent was arrested by a pitch input about 30 s later.

Miami gave the aircraft a new heading of 270. The aircrew acknowledged and turned to the new heading. The flight crew continued to try and free the nosegear position light lens from its retainer, but with no success. The captain again ordered the second officer to descend into the forward electronics bay to visually inspect the gear. This was a minute after the first request. Why he did not comply the first time is unclear.

At 2338:46 (4 min after the initial go-around), Eastern 401 called Miami and requested to continue a little further west to continue to work on the light. Miami granted the request. For the next three min, the two pilots discussed how the nosegear position light lens assembly may have been reinserted incorrectly. During this conversation, a half-second C-chord alarm sounded in the cockpit. The C-chord alarm indicates a deviation of ± 250 ft in altitude. No crew member commented on the alarm, and no flight control input was made to counter the loss of altitude.

The discussion of the faulty nosegear position light lens assembly was interrupted by the second officer popping his head up into the cockpit. It was 2341 (7 min after the initial go-around). The second officer complained that he couldn't see a thing. He had thrown the light switch and gotten nothing, so he was unable to see the two rods he was looking for, or anything else for that matter. There was an Eastern maintenance specialist riding in the forward observer seat. He and the rest of the crew engaged in a discussion of the nosewheel light. Following the conversation, the maintenance specialist joined the second officer in the forward electronics bay.

Meanwhile, the Miami approach controller noticed an altitude reading of 900 ft on Eastern 401 (it should have been 2000 ft). The approach controller then queried the flight, "Eastern, ah, four oh one, how are things comin' along out there?" It was 2341:40. Four seconds later, Eastern 401 replied with, "Okay, we'd like to turn around and come, come back in." Approach Control immediately granted the request and gave 401 a vector to 180. Eastern 401 acknowledged the vector and began a turn. Less than 10 s later the following conversation ensued:

2342:05   The first officer said, "We did something to the altitude." The captain replied, "What?"

2342:07   The first officer asked, "We're still at two thousand, right?" The captain immediately exclaimed, "Hey, what's happening here?"

2342:10   The first of six radio altimeter warning "beep" sounds began; they ceased immediately before the sound of the initial ground impact.

2342:12   Nearly 19 mi west-northwest of Miami International, Eastern Flight 401 crashed into the Florida Everglades in a left bank of 28 degrees. The aircraft was destroyed on impact. (NTSB-AAR-73-14)

How this happened...the autopilot

How did this aircraft accomplish a gradual descent of nearly 2000 ft and nobody noticed it? The answer for the descent seems to be the autopilot system. Realize that in this time frame many new aircraft were being introduced. The Boeing 727 was relatively new, and the 747 was coming on line as well. The DC-10 was being introduced by McDonnell-Douglass, and the L10-11 TriStar was rolled out by Lockheed. These new aircraft possessed new automated systems that were much more reliable than previous systems found on older aircraft such as the Boeing 707 and the Douglas DC-8. Therefore, pilots were just getting accustomed to relying on such automation, and specifically autopilots, at points in the flight other than cruise altitude.

The accident investigation board spent a great deal of time and effort to ascertain exactly who was controlling the aircraft at which point, humans or the autopilot. The investigation centered on determining in exactly what mode the autopilot was engaged and whether the auto throttles were used. (For details of this analysis see the NTSB report.) Through the use of the fight data recorder, cockpit voice recorder, and expert testimony, the board made the following conclusion concerning the descent. The first officer manually flew the aircraft until 2236:04, when the captain ordered the engagement of the autopilot. There were two autopilot systems on this L10-11, A and B. Though it was of interest to the board to determine which of these systems was engaged, this discussion is beyond the scope and purpose of this book. The board determined that System B was engaged with altitude hold and heading select functions in use. The first officer probably selected 2000 ft on the altitude select/alert panel.

At 2337, which was 4 min and 28 s prior to impact, the aircraft recorded a vertical acceleration transient of 0.04 g, causing a 200-ft-per-minute rate of decent. This was most likely caused by the captain turning to direct the second officer to go down into the forward electronics bay to check the condition of the nosewheel. The captain likely turned to talk with the second officer and inadvertently bumped the control column. This bump would be sufficient to disengage the altitude hold mode of the autopilot. When the altitude hold mode was disengaged, it should have been indicated by a change on the anunciator panel (a panel revealing the status of the autopilot with little lights), the extinguishing of the altitude mode select light. There is no way to verify if this light indeed extinguished, but it is clear that the crew did not notice such a change if it was present.

Roughly 30 s after this bump and disengagement, the crew was given a vector of 270. The flight data recorder indicates that this turn was accomplished with the heading select mode of the autopilot. To accomplish this turn, the first officer would have had to act. This would include giving attention to the autopilot control panel to work with the heading select functions. The lateral control guidance system achieved the turn to 270 and commanded a 0.9 degree pitch-up maneuver during the change of heading. This accounts for the pitch-up after a 100-ft descent.

Once the aircraft reached a heading of 270, the gradual descent resumed, followed by a series of small throttle reductions on various engines over the last 2 min 40 s of the flight. The board determined that these reductions were caused by slight bumping of the throttles as the captain and first officer leaned over to work on the jammed nosegear position light assembly. As each individual throttle was moved slightly, the other throttles would be manually moved to even the throttle arrangement. The aircraft never approached stall speed. The descent caused by the inadvertent disengagement of the autopilot altitude hold function accompanied by the throttle reduction (likely causing a slight nose drop) eventually was arrested by ground impact. At impact the aircraft was descending at approximately 3000 fpm. The radio altimeter aural tone did sound at 101 ft (50 ft above the decision height set in the copilot's radio altimeter, which was 51 ft). This was 2 s before impact.

The controller

After understating the situation aboard the aircraft, the first question that many ask is, "What's up with that controller?" Why wasn't he more concerned with this flight and its loss of altitude? The controller later testified he contacted Eastern 401 primarily because it was nearing his boundary of jurisdiction. He was not overly concerned with the plane's altitude because momentary altitude deviations on his radar display were not uncommon. In fact, according to analysis, the ARTS III radar equipment (in use at that time) can indicate incorrect information for up to three sweeps. This, accompanied by the aircraft crew's immediate response that it was fine and heading in for a landing, gave the controller no concern for the flight's safety. Assuming no controller action was necessary, he turned his attention back to the other five aircraft in his airspace.

The nosewheel landing gear

The investigation following the crash was clear. All of the landing gear, including the nosegear, were down and locked when the aircraft crashed. The nosegear down and locked visual indicator sight and the nosegear well service light assembly were both operative. The second officer's failure to be able to see the nosegear indicators through the optical sight could have been because the captain failed to turn on the light in the nosegear well by a switch located on his eyebrow panel. This seems likely, since the crew's conversation inferred that it assumed the light would be on anytime the nosegear was down. It is possible that the captain turned on the light and it was inoperative. It is also possible, but unlikely, that the second officer failed to operate a knob that removes the optical tube cover on the sight. This is a good lesson that the knowledge of one's equipment is imperative, especially in an emergency situation. As Chuck Yeager always said, knowing your machine can be the difference between life and death.

But the bottom line is this: The gear was completely functioning properly. The problem was two burned-out lightbulbs in the nosegear indicator light on the panel. So the problem that precipitated this crash was not a malfunctioning landing gear, but rather two lightbulbs that cost less than a dollar. The nosegear warning light lens assembly that took so much of the crew's attention and effort was found jammed in a position 90 degrees clockwise to and protruding ¼ inch from its normal position.

The captain's brain

Interestingly, the autopsy on the crew revealed that the 55-year-old captain had a brain tumor. The tumor was located in the right side of the tentorium in the cranial cavity. The tumor displaced the adjacent right occipital lobe (responsible for vision) of the brain. The tumor was approximately 4.5 cm (around 1.5 inches) in diameter. The location of the tumor could have affected the captain's peripheral vision. If this was the case, he may have failed to detect changes in the altimeter and vertical velocity indicators (VVIs) as he watched the first officer work with the nosegear light lens. However, the exact effect, if any, on the captain's peripheral vision because of the location of the tumor could not have been predicted with any accuracy. Furthermore, doctors testified that an individual with this type of tumor would have slow onset of peripheral vision loss and be able to compensate for the loss without close associates or even him or herself being aware of the loss. Relatives, close friends, and fellow pilots testified they had witnessed no loss of vision in the captain, as he was able to perform any and all duties requiring peripheral vision. Therefore, the board ruled out the captain's tumor as any sort of cause of the accident. This does remind us, though, of why we have the requirement to pass a Class A physical in order to operate an aircraft. Aviation is a demanding profession and requires that a pilot be physically fit to operate an aircraft safely.

Channelized attention

At numerous points in this book, we have seen the ill effects of channelized attention. When the crew focuses on one aspect of flying to the exclusion of other cues, channelized attention has occurred. Channelized attention leads to lost situational awareness as surely as a river runs to the ocean. The accident board's listing of the probable cause of this crash reads, "Preoccupation with a malfunction of the nose landing gear position indicating system distracted the crew's attention from the instrument and allowed the descent to go unnoticed."

The terms *channelized attention, cognitive tunneling,* and *fixation* are nearly synonymous. However, I think a case can be made that fixation is a more acute stage of channelized attention. Fixation occurs when you focus on only one element of the environment. In this particular instance, the crew spent an inordinate amount of time focusing on the nosegear position light lens assembly. This small assembly normally indicates the gear position. Between trying to determine whether it was functioning properly and trying to dislodge it from a jammed position, the crew members spent much of the last 240 s of their lives fixated on the assembly. This fixation was to the exclusion of all other indicators that could have told the crew it was descending. The lights outside on the ground, the pilot and copilot's altimeter, the pilot and copilot's VVI, the pilot and copilot's radio altimeter, and the pilot and copilot's attitude indicators were all signaling a descent. But these signals were missed because of fixation on the nosegear light assembly.

Poor CRM

Part of CRM is using the resources on board effectively: crew members and equipment. The decision to utilize the autopilot while troubleshooting the landing gear sys-

tem was a good one. Johannsen and Rouse in 1983 studied pilot planning behavior in normal, abnormal (e.g.,
runway temporarily closed because of snow removal),
and emergency (e.g., loss of #2 engine) workload conditions using a HFB-320 Hansa Jet simulator. Not surprisingly, they found a much greater depth of planning
in emergency situations when there was an autopilot
available. The autopilot allowed the pilot to devote
attentional resources more fully to the emergency situation. However, this does not suggest that the pilot stop
flying the aircraft altogether. The failure on the captain's
part to assign someone to monitor the progress of the
aircraft was a poor decision. Remember, the first rule
in any emergency is to maintain aircraft control. A corollary is to keep maintaining aircraft control. The crew did
not notice its drop in altitude until seconds before
impact.

Automation

Flight automation is a wonderful innovation. It has
many benefits. However, it is not a cure-all. Automation
was a problem on several accounts in this accident.
While we don't often think of it as such, the landing
gear system on most aircraft is automated. The pilot simply lowers the gear, and machinery takes over from
there—dropping gear doors, deploying the gear, checking that it is down and locked, sending signals to the
gear indicators in the cockpit, reclosing all or some of
the gear doors, and so on. One problem with automation is that we sometimes don't completely understand
it. The visual check of the nosegear is a good example.
It seems from cockpit discussion that the crew assumed
the nosegear well light would be on whenever the
nosegear was down. This was not the case; the captain
was required to activate a switch on his panel for the

light to properly function. The crew seemed to assume (incorrectly) that the light was automatic.

Ignorance of automation was key in another aspect of the ill-fated flight. The crew seemed to be unaware of the low force gradient input necessary to bump the autopilot out of the altitude hold position when in the CWS (control wheel stering) mode. Upon further investigation, it was ascertained that this was a widespread problem within the company, largely caused by a company policy to only operate the autopilot in command modes that did not include CWS. This is an excellent example of James Reason's higher levels of error prevention failing because of poor organizational policy and guidance.

This accident occurred before the advent of GPWS. But shouldn't the crew have received some kind of notification of a departure from altitude? The first indication the crew actually received was when it reached 1750 ft after departing 2000 ft in a slow descent. Whenever the L10-11 departs a selected altitude by ± 250 ft a C-chord sounds once and an amber warning light flashes continuously. However, on the Eastern Airlines configuration, the light was inhibited from operating below 2500-ft radar altitude. So all the crew received on this flight was the C-chord, which it did not hear. This is an example of poor automation design, which potentially could have prevented a mishap. One of the safety recommendations of this report was to have the amber light flash anytime the aircraft departs an assigned altitude by more than 250 ft, regardless of the aircraft's altitude.

The final bit of automation that was designed to warn the crew of impending disaster was the radio altimeter. As mentioned previously, the L10-11 has a radio altimeter that displays altitude above ground level (AGL). The captain and first officer each had a radio altimeter. The decision height is often set in these altimeters. The captain's radio altimeter was set at 30 ft and the first officer's at 51 ft. The Tri-Star had an aural tone designed to ring 50 ft above the selected decision height. This tone did indeed ring 2 s before impact. There is not any indication that the crew tried to pull the yoke back at this point, or at any time in the last 7 s of the flight when the copilot first noticed something amiss with the altitude. This could be a case of the "deer in the headlights," as the crew froze when it began to perceive the danger signals all around. I may have done the same thing as this crew. But the correct response would be to immediately pull back on the yoke and ask questions later. If we are to use automation effectively, we must know when immediate response is required from automated warnings and when more thoughtful diligence is called for. It is very rare that immediate action is required in a cockpit, but when it is, it must be just that—immediate.

Things probably happened too quickly to expect this crew to react. However, this deer-in-the-headlights phenomenon is sometimes referred to as *resignation*. Another name is the what's-the-use syndrome. It is when further action by the crew seems hopeless to every crew member and so no one acts but rather freezes: "Why act? It's too late; it won't make a difference anyway." The crew members didn't have time to verbalize their thoughts, but one of those thoughts may have flashed through their minds in an instant. This attitude is similar to the one displayed by Eeyore, one of the Winnie the Pooh characters. Eeyore always thinks any idea or response is useless, but it is not! The best remedy for this syndrome is always "have an out in your back pocket." A great example of a crew that could have succumbed to this attitude but didn't was the UAL DC-10 crew that lost all hydraulics and was still able to crash land in Sioux City, Iowa. The crew never gave up, and as a result, many lives were saved.

My final thoughts on this accident, so rich in lessons to be learned, are the prophetic comments made by the accident investi-gation board. Automation was progressing rapidly in this time period. Pilots testified to the board that "dependence on the reliability and capability of the autopilot is actually greater than anticipated in its early design and its certification." The board forecasted that crews would continue to grow more and more reliant on these sophisticated systems to fly the aircraft, especially as the reliability of such systems improved. They also warned that supervision of the aircraft's flight and basic aircraft control would slip as pilots turned their attention to seemingly more pressing tasks and overreliance on automated systems increased. Their words would prove prophetic indeed.

Follow the Bars

"It was a San Francisco International (SFO) takeoff on Runway 19R. We planned the Dumbarton 6 departure. We caught the note on the departure procedure: 'RWYs 19L/R departure turn left due to steeply rising terrain to 2000 ft immediately south of airport.' We were in the Airbus A319 and took off with NAV armed. At 400 ft, NAV engaged; however, the flight director never commanded a turn (I didn't get the steering bars), and I continued to fly straight ahead. The tower issued a warning to turn left to avoid terrain. I selected 'HDG' (which caused the command bars to direct a turn to my slewed heading) and turned. At times we can be complacent with FMS aircraft. At all times I plan to be more diligent concerning situational awareness and not blindly follow unsafe flight director commands." (ASRS 421272)

The FMS will take care of things

Modern aircraft are equipped with a number of items to make flying safer. Many of these tools are designed to relieve the pilot of a high mental workload—in essence, to help make the job of flying easier. A good example is the steering bars that many aircraft have on the attitude indicator, or attitudinal direction indicator (ADI). These bars can help direct the pilot to a course or a heading selected through the FMS or autopilot system or an old-fashioned heading bug on the horizontal situation indicator (HSI). If the pilot simply follows the command bars, the aircraft is directed toward the desired state. The steering bars are also helpful in flying instrument approaches. This is very simple in theory and most of the time simple in practice.

The problem is that the pilot can grow complacent because of his or her reliance on such tools. When the tool fails, we either don't notice or we are caught off guard and it takes a while to regain our situational awareness. This was the case with the SFO pilot. He was waiting for the steering bars to command a turn. When they didn't, he continued to fly straight ahead toward the high terrain. It took controller intervention to warn him of his peril. Notice that the other crew member didn't catch the mistake either. It seems that he or she may also have been lulled into waiting for the bars to work.

We can get used to automation taking care of things. A major problem with automation is that when it fails, the pilot may be so rusty at the basic (unautomated skills) that it is difficult for him or her to intervene appropriately to fix the situation. It is good practice to fly some approaches and takeoffs the old-fashioned way: by simply using the ADI and other instruments without the aid of the automated guidance systems. That way, when and if they fail, you will be ready.

The Frozen Co

The MD-80 was descending into the Spokane Valley near the "Lilac City" of Spokane. While being vectored to

Spokane International Airport (GEG) at 4200 ft MSL on downwind to Runway 21, the crew was given a turn to a heading of 070 degrees and issued a descent to 4000 ft MSL. Bill was the captain and Jim the first officer. Jim had control of the aircraft. He was on his second operating experience-training flight (second flight out of the simulator) with the company. At 4100 ft, the silence was broken when the crew received a GPWS warning of "Terrain, Terrain." The first officer did nothing. Bill seized the controls and executed an escape maneuver to 4500 ft MSL. The warning ceased. After landing, the crew had a discussion with the tower to see if it would pass on its information to Approach. When the crew debriefed, the new first officer said he learned something: GPWS warnings don't allow problem solving by guessing what is wrong— you must react, then figure out what went on after the danger is over. (ASRS 424892)

Hit the nail on the head

A few years ago, Gary Larson, the creator of *The Far Side,* did a cartoon with a captain frantically trying to save a burning aircraft. The captain cried out, "Things are bad and my copilot has frozen up!" Seated next to the captain was a snowman adorned with a airline hat and tie. *The Far Side* brings chuckles and illustrates what happened to this new first officer. He froze up. The captain had to take the aircraft away from the first officer because of his inaction. Automation only works if it is heeded.

Very few things require immediate action on the part of a crew. Most of the time, we must think before we act. However, in the case of a GPWS warning, we must act immediately in accordance with well-established procedures. Most organizations have drilled into crew members the proper procedure to follow in the case of a pull-up warning by the GPWS. The stimulus and response are clear. You hear the alarm, you execute the escape maneuver. However, it seems that this procedure is not always followed. The Flight Safety Foundation's CFIT report found that of the accidents it studied, in 30 percent of the mishaps involving aircraft equipped with GPWS, there was no crew reaction to the GPWS. The first officer hit the nail on the head with what he said above about reacting promptly to GPWS warnings.

A strong captain

This captain did a couple of things that indicate he is an effective leader. First, he took decisive action when his first officer was frozen in inaction. This is excellent CRM. The captain intervened in a timely manner in order to avoid disaster. He recognized the state of his first officer and responded accordingly. A good captain can read his or her people.

Second, the captain sat down with this young first officer and had a thorough debrief. A good postflight analysis is one of our most effective tools in controlling pilot error. The captain could have chosen to minimize the incident and dismiss it with a comment like, "Well, I'll guess you'll know better next time." That is an unprofessional response. This captain analyzed the situation thoroughly, allowing the first officer to arrive at his own conclusion. When we arrive at conclusions ourselves, we are more apt to take ownership of them. Because of the captain's actions, this first officer has a lot better chance of reacting correctly the next time to mitigate inflight error.

Cali, Columbia

On December 20, 1995, an American Airlines Boeing 757-223, Flight 965, was on a regularly scheduled passenger flight to Alfonso Bonilla Aragon International Airport, in Cali, Columbia, originating from Miami International Airport, Florida. Flight 965 was initially delayed for 34 min by the late arrival of connecting passengers and baggage. The flight was then further delayed after leaving the gate for 1 h, 21 min because of gate congestion and airport traffic. The flight finally departed with an estimated time en route of 3 h, 21 min.

The events of the flight were uneventful and of insignificance up to the point where AA965 entered Cali airspace on descent to FL 200 to the Tulua VOR, located approximately 34 mi north of Cali. AA965 contacted Cali Approach Control at 2134. After the initial call the controller requested the aircraft's DME from Cali, and the captain stated that it was 63. The controller then said, "Roger, is cleared to Cali VOR, uh, descend and maintain one-five thousand feet...report, uh, Tulua VOR." The captain then replied, "OK, understood. Cleared direct to Cali VOR. Uh, report Tulua and altitude one-five, that's fifteen thousand three zero...zero..two. Is that all correct, sir?" The controller responded that everything was correct.

Unfortunately, the controller was incorrect and there was already a small misunderstanding between the pilots and the approach controller. "The flight is not cleared direct to Cali. Cali VOR, which is about nine nautical miles south of the airport, is the clearance limit. In fact, the flight is first supposed to report over the Tulua VOR, about 34 nautical miles north of the airport" (Garrison 1997, p. 93).

At 2135 the captain told the first officer that he had "put direct Cali for you in there." At 2136 Cali Approach asked, "Sir, the wind is calm; are you able to approach runway one niner?" The captain then asked the first officer, "Would you like to shoot the 19 straight in?" The first officer replied, "Yeah, we'll have to scramble to get down. We can do it" (Aeronautica Civil, p. 15). The captain responded to the controller with, "Uh, yes sir, we'll need

a lower altitude right away, though." The controller replied with, "Roger, American nine six five is cleared to VOR DME approach runway one niner. ROZO number one, arrival. Report Tulua VOR."

From this point of the flight to impact, the crew is confused as to where it is and how to get to where it is going. The crew is also under a great time constraint as well, as it has changed from its original flight plan, which called for an American Airlines standard: to approach and land on Runway 01. The initial confusion can be seen in the comments between the pilot and first officer in which the captain says, "I gotta get you to Tulua first of all. You, you wanna go right to Cal, er, to Tulua?" The first officer replies, "Uh, I thought he said the ROZO one arrival?" The captain responds, "Yeah, he did. We have time to pull that out(?)...". At this point, 2137:29, the captain calls Cali Approach and requests direct to ROZO and then the ROZO arrival. The approach controller approves the request, and the flight continues.

During this time the captain and first officer are struggling to determine their location. They are having difficulty finding the Tulua VOR (ULQ) on their flight management computer (FMC). What they do not know is that once a new route has been selected, all previous navigation aids are erased from the map and the associated screen. The previous route was erased when the captain entered in the new route, which should have been a direct flight to the ROZO NDB.

When entering information into the Flight Management System (FMS), the first letter of the identifier is entered and a menu of navigation aids is displayed, with those in the closest proximity displayed first. Unfortunately, the FMS displayed a large list of navigation aids north and south of the equator whose identifiers began with the letter R. The captain assumed that the first NDB on the list would be the ROZO NDB as is usually the case. Unfor-tunately, Romeo was actually the first NDB on the list, and it has the same frequency and same Morse code as the ROZO NDB. However, the Romeo NDB is 132 mi north-northeast of Cali in Bogotá. The Romeo NDB listed first as Bogotá is a larger city than Cali and placing it first suits a larger number of users. The captain inadvertently selected this route and entered it into the FMS. The new route caused the aircraft to turn to the left to satisfy the route change. During this turn, the first officer asked, "Uh, where are we...yeah, where are we headed?" The captain replied, "Seventeen seven, ULQ uuuuh, I don't know. What's this ULQ? What the...what happened here?"

At 2137, after passing the Tulua VOR during a maintained descent and turning to the left, the airplane flew on an easterly heading for one minute. The first officer then took over the plane manually and attempted to correct the course to the right. At 2140 the captain confirmed with approach control that he wanted them to go to Tulua VOR and then do the ROZO arrival to Runway 19. The controller answered with a statement difficult to understand and then asked for AA965's altitude and DME from Cali. The captain replied with, "OK, we're thirty seven DME at ten thousand feet." Unfortunately, this put AA965 6 mi past the Tulua VOR, its initial approach point and reporting point.

The captain and first officer proceeded to search for and locate the Tulua VOR, which was well past them. They then decided to fly direct to ROZO instead of dealing with the Tulua VOR. Briefly, the captain challenged their proceeding on to Runway 19 but was ignored by the already busy first officer. Approach control then contacted the aircraft and requested altitude and distance. The captain replied that the plane was at 9000 ft but he was unable to give a distance, as the Ground Proximity Warning System (GPWS) sounded and instructed the crew to pull up.

The flight data recorder (FDR) indicates that the crew applied full power and raised the pitch of the aircraft but failed to retract the spoilers and was given no indication that they were improperly deployed. The aircraft entered into the stick shaker regime, and the nose was consequently lowered and then again raised into the stick shaker regime. The aircraft impacted at 2141:28. The wreckage path and FDR indicate that the plane was on a magnetic heading of 223 degrees, nose up, wings level, as it struck trees at 8900 ft on the east side of Mount El Deluvio. The aircraft continued over a ridge near the summit of 9000 ft and impacted and burned on the west side of the mountain. (Case study and analysis were adopted from an excellent student paper by Chris Bergtholdt.)

Cockpit crew

The pilot in command, age 57, had a current first-class medical certificate and had approximately 13,000 total hours. His total in type was 2260 h and he had flown into Cali, Columbia, 13 times prior to the accident. The first officer, age 39, had approximately 5800 total hours and 2286 total hours in type. He held a current first-class medical certificate and was in good health. The first officer had never flown into Cali before but had flown to other destinations in South America as an internationally qualified B-757/767 first officer.

Cali air traffic controller

The air traffic controller indicated to interviewers that there were not any difficulties in communicating with the AA965 flight crew. However, a second interview revealed that "he would have asked the pilots of AA965 more detailed questions regarding routing and the approach if the pilots had spoken Spanish." The controller also said that "in a non-radar environment, it was unusual for a pilot to request to fly from his present position to the arrival transition. The air traffic controller also stated that the request from the flight to fly direct to the Tulua VOR, when the flight was 38 miles north of Cali, made no sense to him" (Aeronautica Civil 1996, p. 23). The controller further failed to ask the pilots questions as to their request because he did not command a strong knowledge of the English language.

American Airlines training in Latin American operations

American Airlines provides additional ground school training, including a reference guide, to flight crews that may be flying into Latin America. The guide includes hazards that may be encountered in Latin America. Additionally, part of crews' annual CRM training is devoted to Latin American operations. Topics covered include "Warning! Arrivals May Be Hazardous," "They'll [ATC] Forget about You—Know Where You Are!", "When 'Knowing Where You Are' Is Critical," and "How to Determine Terrain Altitude." The training emphasizes good situational awareness and staying ahead of the aircraft.

Analysis

The analysis of the crash includes decision making, situational awareness, automation, and CRM. First, let's look at the crew's decision to accept Runway 19 instead of following the filed approach to Runway 01. The decision was based on time constraints that began with the original delays in Miami. The need to get on the ground can be seen in the captain's statement to the first officer to "keep up the speed in the descent."

Decision making

The flight crew made a joint decision to take the 19 approach instead of 01 in a 4-s exchange. This resulted in the crew having to complete the following list of activities in a short amount of time in order to successfully complete the approach:

1. Locate and remove from its binder the approach plate for Runway 19.

2. Review the approach plate and determine frequencies, headings, altitudes, distances, and the procedure for a missed approach.

3. Enter the new data into the FMC.

4. Compare navigational data with those being displayed in the FMS for accuracy.

5. Recalculate airspeeds, altitudes, and other factors for the new approach.

6. Speed up the descent, as this approach was much closer than the approach for Runway 01.

7. Maintain communications with ATC and monitor the progress of the aircraft on the approach.

Situational awareness

At this point in the flight the pilots would have to be working at a rapid pace to complete the previous steps. They would also have to work together and complete a great deal of communication to pull off the changed approach. Evidence shows, however, that insufficient time was available to complete the aforementioned steps. The aircraft had already passed the initial approach fix at ULQ, and the crew was struggling to determine where it was. Both pilots displayed poor situational awareness with respect to several factors, such as location of navigation aids and fixes, proximity of terrain, and the flight path. The pilots were most likely using what is called *recognition primed decision making*, where an experienced person can match cues with those of previous situations to make a decision (Klein 1993, p. 146). This most likely occurs under a great deal of stress when time is a constraint. At this point, the decision making process and focus became very narrow and selective even though a great deal of factors pointed to starting the approach over. The pilot's narrowing situational awareness and task saturation can be spotted by several factors, including the "inability to adequately review and brief the approach, inability to adhere to the requirement to obtain oral approval from the other pilot before executing a flight path change through the FMS, difficulty in locating the ULQ and ROZO fixes that were critical to conducting the approach, and turning left to fly for over one minute a heading that was approximately a right angle from the published inbound course, while continuing the descent to Cali." (Aeronautica Civil 1996, pp. 31–32).

The pilot and first officer were both busy trying to determine where they were or were trying to enter data into the FMC. The aircraft continued to descend and the pilots broke a cardinal rule: Both failed to monitor the progress of the aircraft and both failed to keep track of their position. Despite not knowing their location, they continued on with the approach, which could have been for two reasons: "the failure to adequately consider the time required to perform the steps needed to execute the approach and reluctance of decision-makers in general to alter a decision once it has been made" (Aeronautica Civil 1996, p. 31).

The captain initially led the crew astray when he changed the flight path in the FMS. "In so doing he removed all fixes between the airplane's present position and Cali, including Tulua, the fix they were to proceed towards" (Aeronautica Civil 1996, p. 34). Neither pilot recognized that the critical Tulua VOR was missing from the FMS and was the reason for their navigation problems.

The crew was also situationally deficient in dealing with the approach controller. Through a number of radio exchanges, the captain misinterpreted several of the requests by the controller. While the captain demonstrated that he "understood the appropriate flight path necessary to execute the approach, his position report contradicted his statement" (Aeronautica Civil 1996, p. 34).

The crew also demonstrated that it lacked knowledge of its proximity to terrain. Several explanations account for this: Cali was not identified as a Latin American airport on a "hit list" of airports, the pilots had become acclimated to the dangers of mountain flying, night visual conditions prevented the crew from seeing the terrain, terrain information was not provided on the FMC, and the first officer relied on the captain's experience in flying into Cali. The crew also suffered from a lack of information on the CDU (control display unit) about the surrounding terrain.

There is sufficient evidence to show that American Airlines provided the crew with sufficient information about the dangers of flying in mountainous terrain and as to poor conditions in Latin America. It is evident that "the pilots of the AA965 became saturated and did not recognize the hazards the airline had warned them about as they were encountered during the accident approach" (Aeronautica Civil 1996, p. 36).

Automation

The Boeing 757 was one of the first aircraft to feature a glass cockpit and a considerable amount of automation. "Either pilot can generate, select, and execute all or part of the flight path from origin to destination through CDU inputs" (Aeronautica Civil 1996, p. 40). This accident is a perfect example of the complacency and the complexity effects. Respectively, they occur when the "use of automation can lead to reduced awareness of the state of the system" and "that increased automation makes straightforward tasks more complex and interdependent" (Ladkin 1996, p. 4).

The first automation-related mistake was made by the captain when, under time-induced distress, he executed a change in the FMS. The captain executed a course to the identifier "R," which he believed to be the ROZO identifier as depicted on the chart. "The pilots could not know without verification with the EHSI display or considerable calculation that instead of selecting ROZO, they had selected Romeo Beacon, located near Bogotá, some 132 miles east-northeast of Cali." "Both beacons had the same radio frequency...and had the same identifier 'R' provided in Morse code on that frequency" (Aeronautica Civil 1996, pp. 41–42). The pilot had made a simple mistake by choosing the first available beacon, which is usually how the system works when in flight.

The other problem associated with this is that ROZO was not even on the FMS generated list. Even though the identifier was *R* for *ROZO*, the entire name had to be entered into the system because of the rules governing the naming of the FMS database. Unfortunately, the crew was not aware of this. This lack of commonality between the chart and FMS was confusing and time-consuming, and it increased pilot workload during a critical phase of flight.

Another automation related problem was that the crew continued to use the FMS even though it was causing the crew confusion and time. The crew's training had indicated to it that the FMS should be turned off when it is causing more confusion than anything else.

The crew did not have an adequate hold on how to use the automation, and in the end it ultimately was hindering the crew's progress. This accident demonstrates that merely informing crews of the hazards of overreliance on automation and advising them to turn off the automation are insufficient and may not affect pilot procedures when it is needed most.

Crew resource management

Throughout the sequence of events that led to the accident, the crew displayed poor CRM, despite a superior training program. It is evident that the CRM skills of the crew were substandard, as neither pilot was able to recognize any of the following: the use of the FMS was confusing and was not helping, neither understood the requirements necessary to complete the approach, several cues were available to suggest a poor decision in changing to Runway 19, numerous parallels existed between the crew's flight and one studied in a CRM course, and the aircraft flight path was not monitored for over 1 min. The evidence indicates that this crew was given background material and information necessary to avoid the accident, but during a stressful situation, the information was not applied, most likely because the critical situation was not recognized.

The crew also displayed poor CRM skills during the GPWS escape maneuver. The GPWS sounded 9 s prior to impact, and the pilots initiated all of the recovery procedures correctly but did not retract the speed brakes (Ladkin). Tests showed that if the crew had retracted the speed brakes, the aircraft would have cleared the ridge. Procedures also call for the captain to leave his or her hand on the brakes while they are extended. Evidence shows that this procedure was not followed and would have saved the aircraft. This accident also demonstrates that even solid CRM programs, as evidenced at AA, cannot assure that under times of stress or high workload, effective CRM will be manifest when it is most critically needed (Ladkin).

The report goes on to list the probable causes of this aircraft accident:

1. The crew's failure to adequately plan and execute the approach and its inadequate use of automation

2. Failure of the crew to terminate the approach despite cues suggesting that it was ill-advised

3. The lack of situational awareness demonstrated by the crew in terms of proximity to terrain, its location, and the location of navigation aids

4. Failure of the crew to revert to radio navigation when the automation became confusing and demanded an excessive amount of attention

The report goes on to make the following recommendations, two of which are germane to this discussion:

1. Require airlines to provide their pilots, during CRM training, with information to recognize when the FMC becomes a hindrance to the successful completion of the flight and when to discontinue the use of the FMC

2. Require airlines to develop a CFIT training program that includes realistic simulator exercises

Error chains

This study is an excellent example of what is known as an *error chain*. An error chain is a series of (seemingly) minor errors that added together cause a catastrophe. If the error chain is broken at any time along the chain, then the accident will be avoided. In this situation, the first error was getting rushed and trying to make up time that was lost at the gate. This led to a decision to take an approach that would save a few minutes. This time compression led to poor situational awareness among the crew members. The poor situational awareness was exacerbated by improper use of automation, specifically the FMS, and an ignorance of its proper functioning. This in turn was worsened by poor CRM among the pilot team, all of which led to CFIT. In the spirit of the DC-9 pilot's advice earlier, if automation is becoming a hindrance to the safe operation of the aircraft, turn it off and do it the old-fashioned way—fly the airplane!

A Final Word on Automation

These case studies have highlighted various problems associated with our helpful friend, automation. Like a diamond, automation has many facets and some flaws. Chris Wickens's work was highlighted earlier in this chapter. He has written extensively on some of the problem associated with automation. Many of the problems he has written about are relevant to the case studies we have reviewed. In order to help get a better handle on these problems, I will review them in turn, illustrating from the previous case studies. The goal is a better understanding of the problem, leading to better prevention for pilots like you and me.

Make sure I operate properly

Have you ever watched someone else play a video game? It is usually very boring. That is how most pilots feel watching an autopilot or FMS control their aircraft...bored. When the human is taken out of the control loop, many times the only thing left to do is to monitor the automation and eat your box lunch, or if you're lucky, the first-class meal. Humans aren't very good at that (monitoring, not eating). The automated systems we have are for the most part very reliable, so we don't expect them to fail. Often our attention is so degraded that when an actual problem arises, we don't notice unless there is a loud alarm. With a great deal of automation the old expression can quickly come true: "hours of boredom punctuated by moments of sheer terror." Several years ago a commercial airliner flying east to west across the continental Untied States bound for California ended up an hour over the Pacific because all of the crew members had fallen asleep in the cabin. Many have commented on the necessity of a balancing act—that is, balancing the workload demands on the pilot so that he or she has enough to keep busy and interested but not so much that the pilot is overwhelmed with information or duties. The balancing point can be hard to find. Barry Kantowitz and Patricia Casper say that our next challenge in automation is insufficient mental demands on the pilot. We are in danger of losing pilots because of boredom.

Complacency

Complacency means letting down your guard and taking things for granted. When it comes to automation, complacency comes nestled in a concept known as the *generation effect*. The generation effect is simply the idea that people remember things better when they choose to do them as opposed to passively watching those choices being made by others, including machines. So here is what can happen: We are poor at monitoring, others (the machine) are making the choices, so we don't update our situational awareness (SA) very well. Once our SA drops, we are slower to detect the system not responding to inputs as it should. We are also slower and/or less accurate in diagnosing what has failed. Finally, we are slower to jump into the control loop to make appropriate manual interventions when they are called for. This is what is known as *out-of-the-loop unfamiliarity* (OOTLUF). A good example of OOTLUF is Flight 401 over the Everglades. The crew did not monitor the autopilot very effectively, so its SA dropped. The crew was slow to diagnose a failure in the altitude hold function. The crew in turn was slow in responding to the failure by manually inputting its own pitch-up via the yoke. The same situation occurred on the San Francisco departure in the Airbus. The steering bars were going "to handle" the departure, including the turn to avoid the high terrain that the crew knew about. The crew was slow to detect the system not responding to inputs. The crew was unsure about it and had to be prompted by the controller to begin the turn. At this point the crew intervened manually and selected another switch to get the job done. Good thing the controller was there. As the pilot said, he needs to be more diligent with his SA and not blindly follow an unsafe flight director.

How does this thing work?

Another problem with automation is complexity. This has been a growing concern on flight decks over the past decade. (In fact, *Aviation Week and Space Technology* devoted an issue to the question "Who's in charge?".) There are two kinds of problems here. The first is that the number of algorithms and hardware that interacts on modern aircraft is mind-boggling. The sheer numbers increase the chance of failure somewhere in the system despite the fact that individual components are getting more reliable. To compound the issue, if a problem does occur, debugging such an intricate system is an uphill battle at best.

The second problem with complexity is more germane to the pilot. Pilots often don't always understand the automation (particularly the FMS). One reason is the aforementioned algorithms, but also these systems use complex logic, logic that is different from yours and mine. In an aircraft, a large variety of actions can be used to reach the same goal. For example, the pilot can change an altitude using five different modes in the FMS. Furthermore, the FMS uses a series of logic functions that may be different for different automation modes. This may put the actions of the FMS beyond the "SA grasp" of the pilot. The pilot's SA may be so out of whack that automated corrections go unnoticed by the pilot. Even worse, the corrections may appear as a surprise to the pilot, who may perceive (falsely) that it is a system failure.

The American Airlines crew flying into Cali illustrated these problems well. The crew members of Flight 965 simply did not understand what the FMS was doing or how to use it properly or effectively. It was over their heads. Their erroneous entries and the resulting erroneous actions of the aircraft compounded an already cloudy situation. All SA was lost as the crew rushed in vain to save a few minutes it had lost in Miami. The complexity of the system caused utter chaos in the cockpit. The crew would have been much better off to shut the thing off and proceed with its traditional instruments. Eastern Airlines Flight 401 also exemplified an ignorance of automated systems in not being aware of how to check the nosewheel gear if primary indications failed. Furthermore, Flight 401 demonstrated an ignorance of the altitude hold function and how easily it could be disengaged by the crew's actions.

Mistrust and trust: Two sides of the same coin

Early in the advent of both the GPWS and TCAS, the "boy who cried wolf" was busy. In their infancy, these systems often sent out false alarms and alerted crews to danger when in fact there was none. The number of these false alarms grew annoying over time. The algorithms simply were not sophisticated enough to determine a real collision trajectory from one that was not. For example, the pilot was about to alter course and the alarm would sound assuming the trajectory would remain straight. The net result of these false alarms was a serious drop in pilot confidence in such systems—a deep-seated mistrust. As a result, pilots began to disregard the signals. In the worst case they would pull the circuit breaker to disable the system and stop the madness. The problem is that true alarms could be disregarded or not treated with the urgency that they demanded. What this would look like is illustrated by our first officer who was flying into Spokane International when he received a GPWS warning. He chose to do nothing but continue his present heading and course...until the captain intervened. The first officer chose to ignore the alarm out of ignorance of what it was really telling him, not out of mistrust. However, the response looks the same for both ignorance and apathy. Fortunately, the number of false alarms has dropped dramatically with newer generations of equipment.

The flip side of the mistrust coin is overtrust or overreliance. Eastern Flight 401 over the Everglades is a prime example. The crew put complete trust and reliance on the autopilot to the complete exclusion of all other indications, save the gear, while it troubleshooted the nosegear indication. The Airbus pilot departing San Francisco did the same thing: He let his SA slip while relying too heavily on the steering bars. The crew in Cali seemed to think that as long as it got the right things punched into the black box, the plane would fly the crew right where it needed to go. Unfortunately, the crew didn't have the time to invest such confidence in the FMS it didn't know that well. The trouble with overtrust is that if we allow automation to do everything for us, it will often do it well and reliably for a long time. But when the automation breaks down (and that time will come), we often find ourselves jumping back in with poor SA, unsure analysis, and rusty flying and thinking skills because the machine had been doing it for us all along.

There have been a couple of proposed solutions to these problems. The first is to continue to develop effective displays that clearly convey what is going on to the pilot in very understandable ways (including pictures). The second recommendation, which was made with both Eastern Flight 401 and American Flight 965, is to continue to emphasize training with the automation. If we are going to use these complex machines, we need indepth training on a recurring basis. The airline training facilities need to challenge their crews to interact with the automation—learn, learn, learn. The training needs to be more than what buttons to push. It needs to cultivate a deeper understanding of what the system is actually doing. Though most pilots are intelligent, they need training on equipment that is not intuitive just as they need training in instrument conditions when their inner ear is telling them they are upside down. But remember this: Never let the automation rob you of those basic flying skills that you started with. That will serve you well when the machines go south.

References and For Further Study

Aeronautica Civil, Republic of Colombia. 1996. *Aircraft Accident Report: Controlled Flight into Terrain American Airlines Flight 965*. Santa Fe de Bogotá.

Aviation Week and Space Technology. January 30, 1995. Automated cockpits: Who's in charge?

Garrison, P. December 1997. Confusion at Cali. *Flying*, 93–96.

Johannsen, G. and W. B. Rouse. 1983. Studies of planning behavior of aircraft pilots in normal, abnormal, and emergency situations. *IEEE Transactions on Systems, Man, and Cybernetics*, SMC-13(3):267–278.

Kantowitz, B. H. and P. A. Casper. 1988. Human workload in aviation. In *Human Factors in Aviation*. Eds. E. L. Wiener and D. C. Nagel. San Diego: Academic Press. pp. 157–187.

Khatwa, R. and A. L. C. Roelen. April–May 1996. An analysis of controlled-flight-into-terrain (CFIT) accidents of commercial operators, 1988 through 1994. *Flight Safety Digest*, 1–45.

Klein, G. A., J. Oransu, R. Calderwood, and C. E. Zsambok, eds. 1993. *Decision Making in Action: Models and Methods*. Norwood, N.J.: Ablex. pp. 138–147

Ladkin, P. May 1996. *Risks Digest*, 18(10).

Reason, J. 1990. *Human Error*. Cambridge, England: Cambridge University Press.

Shappell, S. A. and D. A. Wiegmann. 2000. The human factors analysis and classification system—HFACS. DOT/FAA/AM-00/7.

Simmon, D. A. May–June 1998. Boeing 757 CFIT accident at Cali, Colombia, becomes focus of lessons learned. *Flight Safety Digest*, 1–31.

Wickens, C. D. 1999. Aerospace psychology. In *Human Performance and Ergonomics*. Ed. P. A. Hancock. San Diego: Academic Press. pp. 195–242.

Wickens, C. D. 1999. Cognitive factors in aviation. In *Handbook of Applied Cognition*. Eds. F. Durso, R. Nickerson, R. Schvaneveldt, S. Dumais, D. Lindsay, and M. Chi. New York: Wiley. pp. 247–282.

Wiener, E. 1985. Beyond the sterile cockpit. *Human Factors,* 27(1):75–90.

Wiener, E. L. and D. C. Nagel. 1988. *Human Factors in Aviation.* San Diego: Academic Press.

7

Equipment Problems

We have come a long way in aviation since the Wright Flyer. Orville and Wilbur Wright did not worry about instrumentation and very little about other equipment. They had a few controls to manipulate and an engine to keep running, but other than that they just plain flew. Today we are not so fortunate in some ways. Granted, automation and other equipment have allowed us to fly farther, faster, and longer than the Wright Brothers ever dreamed. Additional equipment has allowed us to fly more safely and efficiently, and in some cases, it has reduced the workload of the pilot. Yet equipment has also made aircraft much more complex and challenging to fly. Perhaps it has taken some of the "fly" out of flying. I remember clearly the first time I sat in the jump seat during a KC-10 flight. The KC-10 is used by the Air Force for aerial refueling and cargo hauling. It is a derivative of the McDonnell Douglas DC-10. It seemed to me that the crew spent the majority of its time flipping switches and dials and programming computers. The pilots did very little hand-flying of the aircraft, yet the flight was very smooth and comfortable as we refueled 10 or 12 fighters. My hair was hardly tussled as I descended the crew steps at the end of the flight. It hit me how far we had come from the daring pilot jumping into his biplane with his scarf thrown over the side. Sometimes the aviation purist wonders if it is worth it all.

Regardless of where you fall on the how-much-equipment-is-too-much debate, there are two things for certain. It a lot of ways, aviation machinery is becoming more reliable and dependable. The other certainty is that regardless of how reliable it is, the equipment does and will break at some point. The focus of this chapter is what happens when it breaks.

Time for Precess

"Our flight was planned from Plattsburgh, New York (PLB), to Napoleon Muni Airport (5B5) in beautiful Napoleon, North Dakota. Before departing in my Cessna 172, I received a weather briefing indicating mixed VFR with rain along the route. Since I was still in training, this was not good news. I thought about it and decided to ask another instrument-rated pilot to go along as safety pilot. It was just beginning to be dusk. The fall colors in upstate New York were beautiful.

"The flight proceeded normally until about 30 mi north of Cambridge VOR. At this point, Approach issued a clearance to 5000 ft. Level at 5000 put us in IMC with moderate turbulence. We continued to press ahead, when slowly the directional gyro began to precess. Shortly, we both noticed this, so I began to use the whiskey compass to estimate my headings. We began the approach. Suddenly, during the approach to Napoleon Muni, Approach Control broke off our approach and issued a climb to 5000 ft and a turn to 270 degrees. We were operating in mountainous terrain, and Approach Control saw that we were deviating off course and toward the mountains. We decided to try it again. A second approach was begun, but I could not steer the proper course because of turbulence and the precessing directional gyro. We broke off this approach and asked for direct vectors to Plattsburgh. The center then gave us 'gyro out' steering vectors to PLB. Over Lake Champlain we broke out into VMC. Relieved, we continued to PLB and landed uneventfully." (ASRS 418299)

The right thing to do

The pilot in this case did a number of right things. First, since he was still in instrument training, he asked a qualified instrument instructor pilot to go along as a safety observer. This was a wise decision. In Kern's airmanship model, this falls under the category of "know yourself and your limitations." A less mature, unprofessional aviator would have elected to press on alone, confident that he or she was tough enough to make it. The professional aviator knows his or her limitations.

As the experienced aviator knows, precession with magnetic instruments is fairly common. They can begin to slew off course, sometimes very slowly, and be difficult to catch. Their onset can be insidious. This pilot and the safety observer seemed to be right on top of the situation...a good catch. Small aircraft often do not have a backup directional gyro. In that case, the pilot is forced to transition to the wet (or whiskey) compass. This is the same compass used by Charles Lindbergh in his historic flight across the Atlantic. It is an old-fashioned magnetic compass suspended in fluid. If you account for magnetic drift (depicted on good aeronautical charts), the whiskey compass can be accurate. The pilot must be careful not to put metal objects (especially large ones) near the compass, as this will result in erroneous readings by the compass. However, most wet compasses are up and away from any item that can compromise their accuracy. Don't put objects near the compass and you should be fine. This pilot's catch of the directional gyro precess was followed by the proper decision to transition to the wet compass. That is how you do it. The pilots then proceeded to a series of bad decisions before getting back on track.

The wrong thing to do

Rely on the wet compass to get you to your destination only if that destination is VMC or it is your only option. Shooting an approach with the aid of the whiskey compass in IMC is not the wisest choice, especially if there are alternates within easy distance with better weather. The sage pilot avoids pushing his luck with his or her wet compass. The reason: It simply is not precise enough to rely on in IMC, especially if you are in an area of high terrain, as were our pilots in this case. Here is a good rule of thumb on when to use the whisky compass: Pretend you are a doctor; the whisky compass is good enough to get you to the patient, but not good enough to allow you to do surgery on that patient.

The pilot also made another mistake. If you lose some valuable navigation aids, tell somebody! That somebody you should tell is the controller. If the controller knows your plight, he or she can help you out by keeping an additional eye on you and your safety. If the controller doesn't know, he or she will give you the same attention as anyone else and that may not be enough in your situation. Of course, this decision depends on your type of aircraft and equipment. If you have several directional gyros on different systems, backed up by INS, DNS (Doppler Navigation System), or GPS, you probably don't need to let the controller know every time one of your gyros burps. In the Air Force there is a checklist for lost navigation aids. When you hit a certain point, you declare "min navaids" with the controller. In this case study, in a single engine aircraft in bad weather, with your only directional gyro gone bad, discretion is the better part of valor. Fess up that you need some help. Let the controller know your status.

Why is it that things seem to break when everything else is going badly as well? In this situation, the weather begins to drop below forecasts. The pilot enters IMC, and then rain, and then turbulence. The time we need our instruments or equipment the worst seems like the time that they go out. Get used to it. I think it is a law of nature. Murphy's Law (if something can go wrong, it will) seems to operate best under adverse conditions. Often the bad weather or circumstances worsen our equipment problem or can even cause them, especially if there is a lot of electrostatic discharge in the environment.

These pilots were very fortunate. They began an approach on the whiskey compass without advising ATC. The controller "saved their bacon" by breaking them off the approach as they headed toward mountainous terrain. Unbelievably, these guys were like kids who couldn't stay away from the cookie jar. They tried another approach (presumably still not advising ATC as to their navaid situation), and the approach was broken off again because of turbulence and the inability to maintain course. The guys were a little overreliant on the controller for their safety. We will study this problem in depth in a subsequent chapter. Suffice it to say that this is not a good idea.

The pilots got back on track when they decided to head back home where conditions were much more favorable. They received "gyro out" vectors back to Plattsburgh by ATC. They should have requested these long ago. Depending on the facility, ATC can sometimes give you gyro out vectors for an approach to the runway.

If that fails, the controller can usually give you gyro out vectors to the nearest VMC. Gyro out vectors are a series of turns dictated by the controller. The controller uses the terms *start turn* and *stop turn* as the pilot accomplishes standard rate turns in the aircraft (this angle varies depending on the type of aircraft, but usually it is about a 30-degree bank turn). Once the pilots returned to VMC, they were good to go. Why no one requested the gyro out vectors sooner is unknown. The good news is that after a serious of questionable decisions by these pilots, they were able to break the error chain that was developing. Because the error chain was broken, they returned safely home VFR with some help from the controller. If these pilots had neglected to break the error chain, we could very well be reading a CFIT versus a CFTT report.

Drifter

It had been a long day. Chad Evans, the captain, and Bill White, the first officer, had found that corporate flying isn't always what it is cracked up to be. This was a 13-h day (so far). Both pilots were hungry; it had been 7 h since they had gotten a bite to eat. Things were compounded for Bill, as he had been unable to sleep the night before. He finally got about 4 h, as best he could guess, all of which resulted in a pounding headache for him. It had been a Little Rock, Orlando, Little Rock round robin; the boss had business to take care of. They were on the home stretch now in their C560 turbojet and descending into Adams Field (LIT) in Little Rock, Arkansas, when they encountered IMC, rime ice, rain, and moderate chop. Other than that, it was a nice day! When Chad turned to a heading with the autopilot direct LIT, the attitude indicator remained in a bank when he rolled out. Cross-checking, Chad noticed the radio magnetic indicators (RMIs, which reveal heading information) were 55 degrees off heading.

He quickly switched to #2 and corrected to course. The problem with the RMIs caused the autopilot and flight director to kick off. Chad continued to have problems with the altitude select and autopilot as he attempted to reengage it. It was a struggle to get it back on. The pilots continued to fly using radar vectors to the approach and were cleared to descend to 2300 ft. Suddenly they looked over and noticed their altitude at 2000 ft MSL. Chad stopped the descent and climbed to 2300 ft MSL. Almost simultaneously, ATC noted the altitude deviation. Chad and Bill were thankful for such backup during a time of flight director problems in the cockpit. (ASRS 424362)

Compounding problems

In a lot of ways, this case is similar to our friends from Plattsburgh we just studied. The primary problem was a directional gyro precessing. However, this crew was fortunate enough to have a backup system. Similarly, this crew faced a series of challenges. First, it was physically behind the power curve. Thirteen hours is a long crew duty day with no backup crew. (Corporate pilots rarely have the luxury of a backup crew.) Fatigue had set in, and when fatigue sets in, often poor judgment follows. Compounding the crew's physical condition was the fact that the first officer had gotten only four hours of sleep the night before. Sometimes this happens; we aren't able to fall asleep. This can be caused by indigestion, a fight with the spouse, a problem weighing on our minds, sick kids. —You name it. It just happens. If it does, let the other pilot know that you may not be yourself that day.

Besides the problems with rest and fatigue, the pilots had not practiced good dietary habits that day. It was nighttime and they hadn't eaten since lunch. Maybe they were holding off for a good meal at home. But this dropped their energy level and ability to concentrate. As pilots we have to resist the urge to put off eating with "I'll catch something later." If nothing else, try to carry in your bag a healthy snack such as peanuts, raisins, trail mix, or even peanut butter crackers. They can give you the pick-me-up you need to keep the energy level healthy.

The pilot's physical and mental state made them very vulnerable. Remember, humans in general are poor monitors. We don't do well in monitoring machines in particular, as we saw in Chap. 6. We lose attention and become distracted while the machine does its thing. In this case, Bill didn't have to monitor a machine. He was monitoring Chad, but not very well. Chad was doing the flying, so Bill's job was to back him up. A good rule of thumb is one person flies and the other worries about the other stuff. Bill wasn't very helpful when it came to the backup.

Which way is up?

Just as Murphy would have it, Chad and Bill hit the bad weather and then their equipment problems hit. The weather doesn't get much worse: IMC, turbulence, rime icing, and rain. Unlike the crew from Plattsburgh, these pilots had multiple heading readouts on board the aircraft. When two systems give you different readings, the best thing to do is to cross-check them with the wet compass. Chad had a good idea which system had failed, as his attitude indicator remained in a bank after he rolled out. He had a good cross-check. Contradictory heading information and frozen attitude indicators can be extremely disorienting. The somatosensory system, the "seat of the pants," tells you one thing, the semicircular canals something else, and the otoliths perhaps something else. The key is to quickly assess and verify which heading system is accurate using your entire array of instruments (VVI, altimeter, and so on), begin using the good system, and ignore the other system.

Once he regained his bearings concerning the heading indicators, Chad turned his attention to the autopilot and altitude select function. It became a struggle to reengage them, and this struggle became time-consuming. Remember the lesson from Chap. 6: If the automation is getting in the way of flying the aircraft, lose the automation. Chad and Bill would have done better here to simply click off the autopilot and manually fly the aircraft down, with one pilot backing up the other.

Back to the channel

The result of all the preceding factors was the now all-too-familiar connection of channelized attention leading to lost situational awareness. This has been a recurring theme in the CFTT and CFIT literature. Both pilots' attention and focus turned to the heading system and the autopilot. Flying took the backseat to these other concerns. Again, we see an error chain developing, a string of little problems leading to potential disaster. Fortunately, the pilots spotted the altitude deviation just as ATC jumped in with excellent backup. Breaking the chain is key to preventing accidents. The remedy for the channelized attention is good CRM, with one pilot flying while the other works the problems with the heading system and/or autopilot. A simple recipe, but one that is hard to follow when fatigued and hungry.

Run Like a Deer

Carlos and Laura were in a C421 on a Friday night in January. Carlos initiated an ILS Runway 34 approach to the Redding (California) Airport (RDD). Carlos was flying as the pilot, and Laura was a commercial-rated instrument instructor pilot, but mostly she was flying as a passenger this night. She had done some of the en route flying and assisted with some of the charts and radio calls. That had helped the 4.5-h air taxi ferry instrument flight to be more tolerable for Carlos. He obtained the current weather from approach control, as the tower and ATIS were closed. The weather called was ceiling 300 ft overcast, visibility 1¾ mi. Carlos also obtained the current altimeter setting and set it in the left altimeter. Redding has pilot-controlled lighting after the tower closes. Carlos believed he activated the lights on approach, but this airport lighting configuration does not provide a means to confirm the status of the lights. He maintained a stabilized ILS approach. The localizer was nearly centered (Laura saw it slightly to the left of center indicating right of course), the aircraft was nearly on glideslope and stabilized. At decision height (200 ft AGL) Carlos saw airport lighting and continued. However, the approach lights (ALSF) were not on. At around 100 ft AGL both pilots felt and heard a bump coming from the left wing, and at the same time, Carlos lost visual contact with the airport and immediately executed a missed approach. At the completion of the missed approach, he received vectors from ATC for another approach. Reaching decision height for the second approach, he could see airport lighting, but still the ALSF lights were not on. Just prior to touchdown, Carlos was able to activate the ALSF lights with five clicks of the CTAF (Common Traffic Advisory Frequency) and landed uneventfully.

During rollout, when he touched the brakes, the aircraft swerved slightly to the right. His left brake was inoperative. He was able to roll to the end of the runway, turn left onto the run-up area for Runway 16, and shut down the engines. Prior to exiting the aircraft, he also noticed a contract security vehicle coming from the area of the localizer antenna. Upon examination, Carlos observed minor damage to the leading edge of the left fiberglass cowling, a small amount of brake fluid leaking from the left landing gear, and a small puncture to the left flap. He then taxied the aircraft to the ramp area. This was further investigated the following morning. It was learned that the left cowling, gear, and flap struck the top of a tree at roughly 50 ft AGL (an approximately 50-ft-tall tree). The tree is located approximately 200 ft left of Runway 34 centerline and approximately 200 ft prior to the runway threshold. The security guard reported that an aircraft hit a deer on the airport within the previous week. Others reported seeing a large herd of deer on the airport that evening. Carlos had personally seen large numbers of deer on the airport in the past. (ASRS 425603)

Who bent the signal?

This is an odd case indeed. An aircraft basically on course, on glideslope during an ILS approach hit a tree. How could this happen? One possibility is that the approach had not been updated for awhile for obstacles. Perhaps there was a fast-growing California fir tree (it can get quite large) that simply grew into an obstacle. While not totally out of the realm of possibility, there is a much better explanation in this case.

First notice that the pilot showed himself on localizer (on course left and right), while Laura showed a slight left deflection indicating he was right of course. Yet the tree was left of course. As Carlos points out in his own analysis of the situation, it is possible that when he broke out of the weather at decision height, he may have mistaken the taxi lights for the runway lights. The taxi lights were basically aligned on a course with the tree. (I personally have known a pilot to land on the taxiway by mistake during daylight visual conditions.) This is easier to avoid at night, as taxiways are lined with blue lights and the runway in white lights. Because of the fatigue of 4.5 h of instrument flying, it is possible that Carlos slid over on the approach once he got the airport lighting in sight. Carlos discounts that he actually made this mistake but does not rule it out as a possibility.

There is a simpler explanation for what occurred. Note when Carlos rolled out after his second approach and landing that a security vehicle was coming from the area of the localizer antenna. There was also a herd of deer seen in the area that night. Indeed, Carlos had seen them on previous visits, and an aircraft had actually struck a deer in the past week. It is very likely that the localizer beam was distorted by an obstruction—the deer, the truck, or both. Normally, a localizer antenna is fenced off to prevent such an occurrence, but deer have been known to leap tall fences. It is not known if this antenna was fenced off or how close a truck could get to it if it was fenced off. What is clear is that wildlife and large metal vehicles can distort our instrument equipment. The pilot here most likely suffered from an equipment failure. Equipment outside of the aircraft, that kind of equipment can get you too.

Turn out the lights

There is one other equipment problem here—that of pilot-controlled lighting. As Carlos mentions, at small airfields after the tower has closed and most everyone has gone home for the evening, there is a lighting system that can be activated by the pilot by clicking on his or her radio switch at a certain frequency. This airport had the ALSF lighting system. The lights are normally kept off to save electrical power until activated by the pilot. The problem with the system is that the pilot is unable to confirm that the lighting system is indeed activated after the clicks. In visual conditions this is not normally a problem, but in IMC it can be the difference between a missed approach and a successful approach. Notice in this situation that Carlos was not able to turn on the lights and confirm they were on until he was almost settled onto the runway during his second approach. If this lighting system had been on during the first approach, he may well have had a better picture of the runway and avoided the tree. He goes on to state that he was so unsure of the status of the lights he may actually have turned them off at one point, when he didn't realize they had been activated in the first place. Until the technology is in place to give current lighting conditions to the pilot in the small aircraft, this is one potential area that the small airplane driver needs to be aware.

Overloaded by Ice

It was a blustery February morning. The air carrier departed Portland (Maine) International Jetport (PWM) in a Beech Turboprop loaded with passengers. During climbout, Mike, the first officer and pilot flying, noticed that the right propeller heat was not working properly. He immediately alerted the captain, Pete. They put their heads together and the decision was made to return to the field. It was a typical day for Maine in the winter; the Beech was in icing conditions with no right propeller heat. The carrier was soon being vectored for the ILS Runway 11. Pete, the captain, was busy talking on the second radio with ops, the passengers, and dispatch. It was one giant coordination exercise. In an effort to get them down quickly, ATC was perhaps too helpful. The vector was somewhat tight, resulting in having to reintercept the localizer after busting through. To help with the workload, Mike selected the flight director, which appeared to give false information. Things didn't seem to add up. Confused and with little time to spare, he selected the flight director to "off." In the midst of this controlled chaos, Approach called and said "1800 ft."

Mike looked down. He was approximately 1600–1500 ft. He immediately corrected back to 1800 ft. (ASRS 427370)

Time, don't run out on me

As has been the case with every equipment malfunction we have looked at, there are many factors at play in this scenario. Again, the weather is bad; it often is in the wintertime in the northern United States and you can't get much more north than Maine. The crew took off and immediately lost right propeller heat, a dangerous condition in a turboprop. The crew made a good decision for a timely return to the field. What it didn't realize is that other equipment malfunctions lay ahead.

As we will see in the next chapter, controllers aren't always perfect. This was part of the problem here. Whether the problem was caused by high traffic volumes, sequencing problems, or a desire to help the malfunctioning aircraft, the controller gave the crew a tight vector. (We have all experienced this, trying to get squeezed in.) The problem with tight vectors is twofold, as we see here. First, it often results in an overshoot in trying to capture the localizer and/or glideslope. The crew is vectored in such a way that a good intercept is not possible. A good intercept is one that requires less than 30 degrees of bank (preferable 15 to 20) to intercept the course. If the crew must use greater than 30 degrees of bank in IMC, spatial disorientation and vestibular illusions are likely to occur. As a result, this pilot had to turn back to reintercept the course. He made the much wiser of the two choices. He could have elected to "bank it up" to intercept the first time or to just turn back to reintercept.

That leads us to the second problem with a tight vector: time. Time gets compressed. The pilot is rushed with the tight vector so he or she may be unable to either intercept the course or accomplish all the necessary checks to get ready for the course. If an overshoot occurs, the pilot is again rushed to reintercept before the glideslope (on a precision approach) or the final approach fix (on a nonprecision approach). When time gets compressed, we get overloaded and we often miss things. In this case, Mike missed his altitude and was heading toward terra firma.

Where was Pete?

Fortunately, Mike had good backup from the controller. But where was the other pilot? That leads us to the next problem area: poor CRM. Yes, in-flight changes require good coordination with the applicable ground parties. But in this case, with so little time to get back on the ground, the captain, Pete, choose a bad time to do that coordination. I remember during the initial qualification course at KC-135 training, they gave us a simulated emergency to talk about around the flight room table. Our dutiful instructors then gave us a list of 12 things that had to be accomplished before we landed. One of those items was "call command post" and another was "eat your box lunch." I'll never forget one old crusty flyer who said, "Call command post as the last step, just after you eat your box lunch." He wasn't totally serious, but we all got a good laugh. There was some truth in what he said. When you have an emergency or in-flight problem, the first step is to maintain aircraft control. Then analyze the situation and take proper action, and land as soon as conditions permit. You'll notice that none of these requires conversation with ops or dispatch. Granted, ops and dispatch can sometimes be of help in troubleshooting a problem and offering suggestions. However, in this case the crew had the problem under control. The communication was to ease things for the arrival back on the ground (e.g., let the gate agent know that passengers would soon be returning to the gate area). Though it would have slowed things once on the ground, this could have waited until landing. In any event, the captain would have been wise to wait for his calls until Mike was established on the approach. If the captain overlooked this, Mike would have been in the right to request that he remain with him until he was established on the approach.

Another part of time compression is high workload. Mike, the first officer, was very busy just flying the aircraft. He had to intercept, reintercept, descend to 1800 ft, and reduce his airspeed. He also had to change gears mentally from "this is a departure" to "this is an arrival." Hopefully, the pilots had briefed an emergency approach return in case this very type of thing occurred. This is an excellent practice used by many pilots. During this period of high workload, Mike made an excellent and timely decision. While trying to reduce his workload, he flipped on the flight director system. When it began to give weird indications (yet another equipment problem), he didn't sink a lot of time into troubleshooting. He simply turned it off and returned to flying the aircraft. He practiced a lesson that the Cali crew in the automation chapter would have benefited from. Unfortunately, the flight director may have been the proverbial straw that broke the camel's back. His instrument scan suffered under the high workload, lack of help from the other pilot, and distraction with the flight director.

When Things Break Down

So, what do we take from our lessons about when things break down? There seem to be several recurring themes in these cases. The first is that equipment problems seem to come in multiples. In three of our four CFTT studies, the crew lost more than one piece of equipment. With increasingly reliable equipment, this should be a rare occurrence. However, sometimes the components interconnect so that a failure of one necessitates a failure of the other. In our second scenario we saw a good example of this, as the autopilot and flight director problems were most likely related to the problem with the radio magnetic indicator (RMI), a heading and navigation system.

It seems that equipment problems happen more frequently in bad weather. Of course, it could be the case that the failures are more salient at those points so we tend to remember them more than failures during good weather. This is known as the *availability heuristic*. Often the bad weather can precipitate the equipment failure. The cold or electrostatic forces can play havoc with electrical equipment and moving parts. Just as our bones tend to ache and creak more in the cold weather, so does an aircraft. Be aware of this higher probability, and you will stand less of a chance of being caught off guard.

If you lose a system with a duplicate or backup system on board, it is imperative to determine which in fact is the good system. Often this will require a deeper knowledge of the system itself and the ability to use other instruments in troubleshooting the guilty party. We should know more about our instruments than just how they turn on and off. The great pilot Chuck Yeager was an aircraft maintenance man before becoming a pilot. This background served him greatly as he went on to shoot down enemy aircraft in World War II and later as he tested many aircraft, including the Bell X-1 that broke the sound barrier. He was almost fanatical about understanding all he could about the X-1 before he flew it. He considered it a matter of life and death. Even though he had only a high school diploma, as far as formal education was concerned, he was a self-learner when it came to his aircraft. He asked the maintenance guys and the engineers every "what if" question he could think of. He would get his buddy to explain the answer to him if he didn't understand it the first time. We would be wise to follow his example, especially as aircraft equipment becomes more sophisticated and can exert its own control over the pilot.

Remember, too, that equipment can fail outside of the aircraft, such as the localizer equipment at Redding, California. A herd of deer or a maintenance truck can affect your flight. There is not much you can do other than fly the approach to the best of your ability, begin to question when things don't add up, and go around if in doubt. Other things are outside of your control as well. For example, know how to remotely control runway lighting the best you can, but realize that you may have no firm knowledge of their status until you break out of the weather.

The condition of your human equipment makes a difference as well. Take care of your body with good nutrition. Carry along a healthy snack, for those times will come when you can't make it to the terminal snack bar. If you are run down and fatigued, try not to fly, but if you must, let your other pilot know. Besides the knowledge of your equipment, your knowledge of one another can make the difference. Good CRM between crew members can compensate for equipment failures. Poor CRM can make equipment failures worse.

Finally, if you have lost some vital equipment, tell ATC. They may be able to help you out. They can make a difference. Their job is to conduct aircraft safely from one sector to another; they usually want to help. But keep in mind they aren't always perfect, as we will soon see.

References and For Further Study

Howard, I. P. 1982. *Human Visual Orientation.*
New York: Wiley.

Yeager, C. and L. Janis. 1985. *Yeager: An Autobiography.*
New York: Bantam Books. pp. 77–114.

8

Controller Error

"If a human being is involved, it will be messed up in some way." Many people adhere to this creed. When it comes to air traffic controllers, we must realize they are not perfect and they make mistakes. Mistakes have different consequences in life, some large, some small. When it comes to ATC the chances that mistakes will have big consequences are significantly higher because of the nature of the business. Controllers are responsible for the safe conduct of aircraft from one sector to another as efficiently as possible. Safety and efficiency aren't always congruent. That is why the controller (and the pilot) should err on the side of safety. Often during flight, the pilots have either no visible horizon or very little visibility at all out of their windscreen. Controllers bear a tremendous responsibility to those aircraft and the lives aboard. The pilots must place their faith in the controller and their instruments. Most of the time both of these entities prove themselves worthy of such faith. Sometimes they do not.

Never Knew the Difference

"I was the controller on duty that day. It was a Saturday morning in November. It was a nice day for the Bay Area. We had a broken deck at 4500 MSL. I was in control of a small aircraft on a pleasure flight. The aircraft departed Reid-Hillview Santa Clara County Airport (RHV) in San Jose, California, and headed to Oakland International Airport (OAK). This would be a relatively short flight of about 40 mi. These are common in the Bay Area. Some folks even commute to work in small aircraft; it beats the automotive traffic! The aircraft departed, climbing to 3000 feet MSL. The aircraft was vectored north of the San Jose VOR (SJC) on a heading of 320 degrees. This heading put the aircraft in a sector with a 4000-ft minimum vectoring altitude (MVA). When I discovered the error, the aircraft was turned to 270 degrees and climbed to 4000 ft. One mile later, the aircraft was turned right again to heading 320 degrees and changed to Richmond sector frequency. Aircraft cross this area all day at 1500 ft and 2500 ft via landmark hotel direct SJC when VFR conditions allow. I feel that the aircraft was in no danger of hitting any mountains. The pilot was on a heading away from hills during the time under the MVA. I wonder if the pilot ever knew the difference." (ASRS 421074)

Consolation

The controller seemed to take some consolation that the error in vectoring the aircraft under the MVA did not seem to endanger the aircraft. The aircraft was headed in a direction away from the high terrain, which dictated the 4000 MVA. The aircraft was 1000 ft below this minimum altitude. I probably would have felt the same way as the controller, but consider this: The aircraft just as easily could have been headed toward the high terrain. An assigned altitude below the MVA toward the hills is a recipe for disaster. What is a pilot to do? Did the pilot even know that he or she was in danger? We will consider that question in a moment, but first we need to examine a related case.

The Western Slope of Colorado

It was January 5, 1999, on a descent to Eagle County Regional Airport (EGE) near Gypsum, Colorado. It is a pretty part of the state on the western slope, not too far from Aspen. Patty was acting as pilot in command of a corporate Learjet 60, operating under part 91 rules. Kevin was in the right seat. During the descent, they received a clearance to descend to 11,000 ft MSL. Kevin read back and set 11,000 ft in the altitude alert system. A few minutes later, they were subsequently handed off to another controller. They descended through about 17,500 ft during the handoff. Kevin checked onto the new frequency "descending through 17,000 ft for 11,000 ft direct RLG" (Kremling VOR). Descending through around 13,000 ft, the controller queried the Lear about its altitude, and the crew responded, "Through 12,600 ft for 11,000 ft." The controller asked, "What are your flight conditions?" Kevin responded, "Just approaching the tops of the clouds." The controller replied, "You were cleared to 14,000 ft!" Kevin replied, "We were previously cleared to and read back 11,000 ft!" The controller replied, "No, you weren't." Enough argument. Kevin replied, "Out of 12,600 ft, climbing 14,000 ft" as Patty pushed up the power and raised the nose. The flight proceeded without further incident, and the aircraft settled onto the tarmac at Eagle County. (ASRS 425310)

Know your altitude

In her report, Patty goes on to make some key points about this incident. "During the descent an inappropriate altitude was given and looking back, should not have been accepted. We had briefed the approach and discussed a minimum safe altitude (MSA) of 15,200 ft. We were given an altitude of 11,000 ft and quite frankly, because of the VMC conditions, it did not set off an alarm in my head as it should have! Had I been following my briefing of the approach, not only would I have questioned the 11,000 ft altitude, but I should have questioned the 14,000 ft altitude as well—whether VMC or IMC conditions prevail!"

Not just one but two bad altitudes were missed in this scenario. One of the key weapons the pilot has to fight CFIT is a knowledge of the local terrain and minimum safe altitudes. Armed with this information, the pilot is able to catch erroneous altitude assignments given by ATC. Without this knowledge, it may become a scenario of the blind leading the blind. In the case of the San Jose to Oakland pilot we first looked at, it is unclear whether the pilot had any idea that the controller had vectored him or her under the MVA. If he or she did not know, ignorance is not always bliss. Granted, the vector was away from the hills, but the pilot may not always know that. In the second case, the crew knew and had briefed that there was an MSA of 15,200 MSL in their area. Yet on two different occasions, the crew accepted altitude assignments below that MSA without a query of the controller. This was because of a memory lapse on the part of both pilots and a false sense of security because of the VMC.

Error types

There are several different types of errors. There are also many classification systems (or taxonomies) for errors. One taxonomy that I find helpful is as follows: Errors of omission, repetition, insertion, and substitution. Errors of substitution occur when a pilot substitutes an action from an old type of system or aircraft into the new aircraft. An excellent example is from the Lockheed F-104 Starfighter flown by the USAF in the 1960s. Originally, the aircraft was designed so that the ejection system ejected the pilot down out of the cockpit, through the bottom of the aircraft. So if a pilot was to eject, he would roll the aircraft on its belly and pull the ejection handles so that he went skyward. As ejection systems became more powerful, the design of the F-104 was changed so that the seat would go up out of the cockpit. However, habit patterns had been established by the older pilots, and you can imagine a pilot's surprise when he rolled the aircraft on its belly and ejected...down toward the earth. This was a poor engineering decision and an illustration of an error of substitution.

Errors of insertion occur when the aircrew does things in the wrong sequence. A young copilot running the After Refueling checklist before the last aircraft has left the refueling boom on the KC-135 is a good example. These errors are often caused by overeagerness or aggressiveness.

Errors of repetition are closely related to channelized attention. A sequence of checks is run repetitively and precludes other vital issues from being attended to. A classic example is Flight 173 into Portland that we examined earlier. The captain was obsessed with repeated gear checks, and the aircraft crash-landed because of fuel starvation.

Finally, there are errors of omission. These involve simply forgetting to do or remember something. You forget even when you know you need to do it. This is the type of error illustrated here. Patty and Kevin had discussed the MSA, but when it came time to put it into practice, they simply forgot about it. This is an error of omission.

Distraction and complacency

How could they forget? They were distracted by the VMC. It was a nice day and they let down their guard. You could say they became complacent. We have already touched on the fact that complacency often leads to a loss of situational awareness. It can also lead to forgetfulness. Vigilance in the cockpit is essential. It is probably even more important when things seem to be fine. If we already sense a tough situation, we will automatically key ourselves up and increase vigilance. So vigilance in good conditions requires good old-fashioned self-discipline. None of us are immune to complacency.

Types of altitudes

We have discussed a scenario where MVA was the issue. Our scenario with Kevin and Patty concerned MSA. Earlier I said that one of the key weapons the pilot has to fight CFIT is a knowledge of the local terrain and minimum safe altitudes. But what exactly are all the different types of altitudes of which the pilot should be aware? There can seem to be a lot of them at times, and it may be difficult to keep them all straight. Following is a list that can perhaps help serve as a reference for those altitudes.

MVA (minimum vectoring altitude) This altitude is not depicted to the pilot. It is the lowest altitude at which an IFR aircraft will be vectored by a radar controller, except for radar approaches, departures, and missed approaches. The altitude meets IFR obstacle clearance criteria. It may be lower than the MEA along an airway or a jet-route segment. Charts depicting minimum vectoring altitudes are normally available only to controllers.

Directional MEA (minimum en route altitude) This is the lowest altitude between radio fixes that meets obstacle clearance requirements and assures acceptable navigational signal coverage.

MCA (minimum crossing altitude) This is the lowest altitude at certain fixes that an aircraft must cross when proceeding in the direction of a higher minimum en route altitude.

MEA (minimum en route altitude) This altitude provides at least 1000-ft clearance over the terrain and 2000 ft in mountainous terrain. Mountainous terrain includes all terrain at or above 3000 ft MSL. The MEA is based on the aircraft being within 4.2 nautical miles (nmi) of centerline.

MOCA (minimum obstruction clearance altitude) This is the lowest altitude on VOR airways, off-airway routes, or route segments that meets obstacle clearance for the entire route segment and assures navigational signal coverage for only 22 nm of a VOR. Shown as a "T" along with an altitude (e.g., 5500T).

MORA (minimum off-route altitude) Known as a Jeppesen altitude, a directional MORA is 10 nm each side of airway centerline and uses an "a" (e.g., 3000a) to distinguish it from other altitudes. A grid MORA clears all terrain or obstacles by 1000 ft when the highest point is 5000 ft MSL or lower and by 2000 ft when the highest point is 5001 ft MSL or higher. Grids are bounded by latitude and longitude lines and are light blue on high-altitude charts and light green on low-altitude charts. These altitudes assume you are within 10 nm of centerline.

AMA (area minimum altitude) This is located on area charts only. It is also known as a Jeppesen Altitude, with the same values as a MORA, but shown only on an area chart. Areas are different shades of green.

MSA (minimum safe altitude) This altitude provides 1000-ft obstacle clearance within 25 nm (unless another distance is stated) from the navigational facility upon which the MSA is predicated. MSA is shown in a circle. When divided, the dividing lines are magnetic bearings toward the facility (a circle shown on heading of Jeppesen approach charts and military charts). The MSA is based on the fix depicted on the approach plate.

In the Flight Safety Foundation's CFIT study, it found that 35 percent of the CFIT accidents examined involved an aircraft descending below some sort of minimum altitude. Besides a knowledge of the minimum safe altitudes on your route and arrival, the pilot has other stores of knowledge that can help prevent CFIT, as we will see in our next case.

Beautiful Palm Springs

"On January 5, 1999, I was pilot in command of a Hawker. I was designated pilot not flying, and my copilot occupied the left seat and was flying the aircraft. We were on an IFR flight plan from Oakland to Bermuda Dunes Airport (UDD), 13 mi south of Palm Springs (PSP), home to the rich and famous. We were flying a corporate jet, but we were 'deadheading' at the time. We had no passengers on board. We had crossed the mountains north of PSP at 14,000 ft MSL and had been given clearance to descend to 5000 ft MSL. It was a Friday morning, VFR, clear with approximately 30 mi of flight visibility. At approximately the PSP VOR, we were given a vector heading of 090 degrees by Palm Springs Approach, which the copilot flew. At approximately 1057, Palm Springs ATC was inundated with VFR traffic calls, which he attempted to respond to. On our heading, the TCAS II system provided two traffic advisories, which we visually identified and avoided (both small piston-engine aircraft). Neither conflict required a heading or altitude change, but there were no calls from ATC.

"Approximately 4 min passed with no further communication to our aircraft, although other aircraft were being provided traffic calls (all apparently small VFR airplanes). Our vector put us well below the ridgeline of a range of mountains east of Palm Springs, which we could see approaching. With approximately 20 s remaining to ground impact, the area to the right visually and TCAS II clear, I instructed the pilot flying to turn right immediately (which he did).

"There were no breaks in the frequency to contact ATC some 2 min prior to our turn, and I used the ident button continuously in the last 60 s before our turn to attempt to get ATC's attention. During ATC's transmissions 15 s prior to our turn, I could distinctly hear an alarm in the background. I am certain this was either a ground proximity alarm or a traffic advisory alarm, although I cannot claim that it was specific to our aircraft. Twenty to thirty seconds after the turn, I was able to break into the frequency and inform ATC we were no longer on his vector 'toward the mountains,' that the airport (Bermuda Dunes) was in sight, and we were canceling IFR and would proceed to the airport VFR. His response, 'report canceling on this frequency or via phone after landing.' I reaffirmed our cancellation, and we landed without further incident at 1104.

"It is clear that the controller lost situational awareness in separation and sequencing of IFR targets under the traffic load. His response indicates that he forgot about our target and it took a moment for him to process any information about us. Although we did not wish to create additional conflicts by acting contrary to our clearance (i.e., assigned heading), it became clear that unless we acted quickly the controller was not going to provide an amended clearance. A contributing factor was the clear weather in the valley, which prompted many pilots to fly this day. I see two possible ways to have avoided this incident: (1) Cancel IFR farther out; this allows the pilot to maneuver for the airport, and if VFR traffic advisories are requested, they can be provided without the constraints of having to follow a clearance. (2) Limit controller workload; each controller must determine a level for which he or she is comfortable working. If the IFR traffic can't be handled properly, then obviously no additional VFR traffic should be accepted for advisories. Supervision is a must to ensure that if the sector gets busy suddenly, the controller does not get overloaded too quickly. I do not think that this case would have happened under IFR conditions for two reasons: (1) The number of aircraft would have been greatly reduced, and (2) our routing would have taken us south in the valley to the Thermal VOR and reversed course for the VOR approach into Bermuda Dunes. As a result of maintaining IFR routes and altitudes, terrain separation would have been provided." (ASRS 425601)

Epilogue

Following the incident the pilot briefed his company chief pilot. Additionally, a few day after the incident, the pilot called the tower to talk with the supervisor. It turns out that the supervisor had witnessed the event unfold from the tracab. He "pulled" the controller in training along with the tapes. He said that both he and the controller had become distracted by VFR traffic calls. Apparently, Palm Springs is a training facility for new controllers and a lot of folks were in training. The supervisor ordered a facility-wide review of instructor training requirements, instructor/trainee scheduling, and qualifications.

We all train

No one initially arrives in a pilot seat fully qualified and proficient. We each go through a series of training programs as we progress from basic pilot to instrument-rated to instructor, and so on. We need to realize that the man or woman on the other side of the microphone does the same thing. Until recently, the FAA had divided radar facilities into five levels of complexity, level 5 being the busiest and heaviest operations (e.g., Terminal Approach Control at Denver International or Chicago O'Hare). Controllers would progress within and then up levels as they become more experienced and proficient. This doesn't occur overnight. In this case, we have Palm Springs, which is a training facility for initial trainees, "the guys right off the street."

The inexperience of the controller was evident in how he prioritized his traffic. Aircraft on an IFR clearance should have priority over VFR traffic receiving courtesy advisories. This priority was not afforded to the Hawker. It seems the controller got distracted by the VFR radio calls on this beautiful and busy day. The controller lost his situational awareness. This distraction led him to forget about the Hawker. Again we see distraction leading to an error of omission, just like Patty and Kevin flying into Eagle County on the western slope of Colorado discussed previously.

If you have ever flown into or around the Los Angeles basin at peak traffic times, you know that the controller often does not pause between commands to a myriad of aircraft. There is no time to acknowledge, as they are speaking nonstop. The Hawker did the appropriate thing in trying to get the controller's attention with both a radio call (he was unable to because of radio chatter) and then the ident button. If that fails and action is required, take the action and talk later! The Hawker did so very professionally. This was right out of the textbook: He was aware of his surroundings (good situational awareness), tried to notify the controller, cleared both visually and with TCAS (using automation wisely and appropriately), and then took evasive action. When the radios cleared, he took immediate action to notify the controller of what had just occurred. The Hawker couldn't have handled the situation any better, which is a good lesson for you and me.

Familiar and unfamiliar territory

There is one other lesson I would like to highlight from the Hawker. The Hawker's crew was aware of its surroundings. This could have occurred because of a thorough study of the local area charts and approach plates during mission planning prior to takeoff. It could have come through conversations with other pilots who had traveled to this airport on previous occasions. Or it could have come from a firsthand knowledge of the local area from personal experience. From their basing information, the latter seems probable. However, all the items I just listed can help you prepare for a flight into a strange area.

Flying into unfamiliar territory can be very disorienting. New areas, with their unknowns and surprises, can cause a pilot to get out of his or her routines and habits. It is sort of like when you first move into a new house and you can't remember where you decided to put the spare rolls of toilet paper or the spare towels. It's also like getting up in the morning and someone moved your razor, and now your routine is all out of whack. The morning just doesn't go as smoothly as it should. Some call this *positional disorientation*, and it can be deadly in an aircraft. Flight into unfamiliar territory is listed as a key variable in causing CFIT according to the Flight Safety Foundation.

To lessen the odds against you in unfamiliar territory, follow the steps outlined above. Thoroughly study area charts and approach plates for high terrain, obstacles, and other hazards. In recent years more information concerning airfields has become available via the Internet. Web sites such as Monkeyplan.com can relay excellent information on almost any airfield worldwide. Be sure to check the old-fashioned Notices to Airmen (NOTAMs) at the departing station for any last-minute changes in obstacles (e.g., crane located ½ mi left of final for Runway 17). Talk to someone who has flown to this location before and pick his or her brain. If you can't find anyone who has been there, consider calling the tower or airfield and ask to speak to someone who knows the airfield and local area. The phone is better than nothing. Once you have gathered the information, make sure you brief it so that every crew member understands what is out there. As a good rule of thumb, brief the high terrain (highest point and altitude) along your route and the highest obstacle on the approach plates to include minimum safe altitudes.

I remember the first time I was to fly to Andersen Air Base in Guam. We pulled aside a couple of experienced pilots to learn about the airfield. The airfield was "loaded with goodies," with high terrain on one end and a sheer cliff on the other with an undulating runway at its center. Their advice was invaluable when we broke through the scattered deck of clouds on our first approach to Guam a few days later. We were prepared; you should be too. The pilot's knowledge of the area in this situation helped prevent catastrophe.

This pilot does a nice job in analyzing the situation, and points out that this scenario would be unlikely under IFR conditions for a couple of reasons. This is probably true; however, in the Flight Safety Foundation's CFIT study, 13 percent of the accidents did occur during VMC. Furthermore, 46 percent occurred during daylight conditions. Controllers can try to send you into the mountains in VMC or IMC. The take-home lesson: VMC doesn't guarantee error-free ATC. The next study we cover illustrates and summarizes the concepts of knowing the territory and the appropriate altitudes well.

Too Many Planes, Too Little Time

On January 8, 1999, Stephanie filed an IFR flight plan from NAS Whidbey Island (NUW) to SeaTac International Airport (SEA) via radar vectors. Her Cessna 172 lifted off, and after takeoff, she flew heading 064 degrees out of the Puget Sound and climbed to 2000 ft, where she radioed NUW Tower and told it she was level 2000 ft. Tower told her to contact Approach. She switched frequencies and told Approach, "Cessna with you, level 2000 ft, IFR to SeaTac." She heard Approach reply, and it issued her the local altimeter and told the Cessna to ident. Stephanie idented her transponder and continued heading 064 degrees at 2000 ft. After what she believed was 7 to 10 nm east of NUW, she radioed Approach with an altitude request. The Cessna received no reply from Approach. Though she received no reply, the airwaves were filled with chatter. The controller was busy with many aircraft. Stephanie continued east, and at about 17 nm from NUW, the pilot flying with her, Andrew, who was familiar with the MVAs in the area, informed her they were in an MVA of 5300 ft. He contacted Approach and advised it that he and Stephanie were IFR and approaching higher terrain and requested an immediate climb and turn to the south. Approach did not reply. After 10 to 15 s, Stephanie began a turn to heading 180 degrees and a climb to 3000 ft. Andrew, the pilot in the right seat, advised Approach that they were turning 180 degrees and climbing to 3000 ft. Approach, after a controller change, replied, "Turn right to 220 degrees, climb and maintain 5300 ft." Stephanie replied, "right 220 degrees and 5300 ft."

After the climb and heading were established, Approach told the Cessna to expedite its climb through 3000 ft. Stephanie complied. She and Andrew were on an IFR flight plan and were not provided IFR separation or radar vectors. The pilots did not believe Approach knew they were an IFR flight until they told it they were turning to 180 degrees and climbing to 3000 ft. They cleared and transversed the area with no other occurrences. (ASRS 424776)

It's all in your expectations

Stephanie and Andrew, who filed the report, make the following observations concerning the flight. At no time did they believe that they were lost com. Stephanie was in communication with the tower and heard Approach give her an altimeter setting and an ident request. There are a lot of light civil aircraft that depart out of NUW. Most of these flights are conducted under VFR flight plans. Stephanie believes that Approach thought she and Andrew were a VFR flight. The ATC is also a training facility for the Navy, and she believes that the controller was under instruction. His traffic load was quite heavy; they heard him controlling many aircraft. To help prevent this occurrence again, she thinks that whenever ATC receives an aircraft, the passing ATC should inform the receiving ATC whether the flight is IFR or VFR, and also that better attention to all flight strips be observed.

Stephanie and Andrew hit on some key points. Let's take them in order. First, Andrew was practicing excellent CRM as the copilot. He could have sat there fat, dumb, and happy. But he chose to be an active participant, sharing important information to help the flight go more smoothly. Notice how he used his knowledge of the local area (familiarity with the territory) to immediately recall the MVA of the area. The MVA was 5300, he knew it, and they were under it. He was proactive with the knowledge.

In Chap. 4 we covered a concept known as *perceptual sets*. These are also known as *mental sets* and are at play in this situation. With a mental set you are primed like an old well pump to expect a certain scenario. You are so sure of your expectation that you don't notice when things are not as they should be. You miss the little cues because of your expectations. It takes something significant to snap you out of it, to get your attention. The controller was very likely under a mental set concerning Stephanie's Cessna 172. Notice she says that there is a lot of light civil aircraft traffic (like her) out of Whidbey Island and most of this traffic is VFR (unlike her). Her call to the controller most likely registered in his brain as "same old, same old...yeah yeah yeah." But this wasn't the same old thing; this was an IFR flight. As the old saying goes, perceptions are real in their consequences. He perceived the Cessna as a VFR flight and treated it accordingly: with little attention. Remember, VFR traffic is responsible for its own terrain separation.

Training and attention to detail

Compounding the mental set is something we covered earlier, so I won't belabor the point. Just like Palm Springs (except much cooler), Navy Whidbey is a training ground for new controllers, mostly military. Inexperienced operators (including pilots and controllers) are much more likely to have a performance degradation under high workload. We certainly witnessed a dropoff in performance in this case. Remember, controllers go through training too.

Finally, we have a case of a lack of attention to detail. Controllers receive a computerized paper flight strip printout on all IFR traffic they are receiving upon takeoff. The strips are not large, nor are they printed in large type. It requires attention to detail to keep up with them, especially if you have a lot of traffic to keep straight. Of course, the controller's screen would also contain valuable information on the flight, if he was paying attention to that Cessna blip. The military stresses attention to detail. I'm sure it was stressed to this new military trainee, once his supervisor got a hold of it.

Did I Mention the New ATIS?

Randy and Stu had copied the 1751 Zulu ATIS information for San Diego International Lindbergh Field, which stated the localizer Runway 27 approach was in use. Aboard the MD-80, Randy, the first officer, briefed, as this was his leg. There was no mention on ATIS of the DME being out of service (OTS). When the pilots were handed off to Southern Cal Approach, they reported they had ATIS information Delta. They were eventually cleared for the approach. They began their descent based on the DME they were receiving. At REEBO, the final approach fix (FAF), they set off the low-altitude alert because they were lower than the published altitude. Surprised, they corrected and made an uneventful landing at 1839 Zulu. They were not advised the ATIS was updated or that the DME was OTS. (ASRS 420460)

Missing letters

What we witnessed here is an error of omission that potentially could have been avoided by the controller and the pilots. When a crew checks in with Approach Control, it is standard procedure to report the ATIS information by its phonetic alphabet designation. ATIS relays information concerning current weather conditions, airfield advisories, and equipment outages at the reporting airfield. It saves ATC valuable time from having to repeat that information to each and every aircraft heading for that field. It is broadcast continuously on a radio frequency depicted on the approach plate, among other places. It is usually updated hourly, with each update receiving a new letter designation proceeding in alphabetical order and beginning again with Alpha after Zulu has been reached. It is often updated around 55 past the hour for consistency. It may be updated more than hourly if conditions are changing rapidly.

Once the aircrew reports in with the ATIS information, it is the controller's job to check it for accuracy. If the ATIS is old, the controller can direct the crew to update the ATIS information or he or she can provide any changes to the crew verbally. It seems likely due to inattention or distraction that this controller was unaware that the crew reported an old ATIS or he was unconcerned about the change, not realizing it may have ramifications for this crew.

Is there anything the crew could have done to prevent this situation? The answer is maybe. A good technique is to listen to the Zulu time posted on the ATIS information. It will reveal what time the message was first posted. If it is old (i.e., approaching 1 h old), the crew should expect an update soon. Of course, this assumes that there is time permitting for a crew member to recheck the status later. This in turn assumes you are not trying to get the ATIS information at the last minute as you rush down on your descent. Always try to get a good feel for the airfield and weather before you begin your en route descent or published penetration. This gives you time to not only "digest" the information but to check back if you think it will be changing soon because of the age of the information or if you suspect rapidly changing conditions. In this case, the crew reported getting the information at 1751 Zulu. It is very likely that the information was updated 4 min later. We don't know if this idea crossed the crew's mind, or if the crew did not have time to update the ATIS information. Either way, it was possible for the crew to have caught the update itself. A vigilant controller would have caught the error, but we should fly in a manner in which we are not overreliant on the controller.

The ups and downs of automation

As a footnote, we see another helpful use of automation in this scenario. The crew reported getting a low-altitude alert at the FAF. It is not clear if this alert was from ATC or from the GPWS. Most likely it was a controller alert and advisory, as there is no mention of an alarming "pull-up" message with accompanying climb. In either case, this is an excellent example of how automation on the one hand caused problems (ATIS information outdated) while being compensated for by another piece of automation (ATC altitude alert). Automation can be a friend or foe.

Just Take the Visual Approach

The eastern Florida coast was glistening that Tuesday evening. The lights of Jacksonville sparkled, adorned with the gem of its new football stadium, home of the Jaguars. Dom and Ryan approached Jacksonville International in their Boeing 737-500 loaded with passengers. Weather that evening included calm winds, a high dew point, and previous thunderstorms in the area. From 15 nm out they could see the airport and runway with no visual obstructions. They accepted the visual approach with airport in sight. They had the ILS to Runway 07 tuned and idented for a backup to the visual. As they neared the runway, the approach end (arrival area) began to fade away (the approach end was an extreme black hole with no surrounding ground lights to help visualize surrounding fog). At approximately 500 ft MSL, they entered a fog bank, "soup," for a few seconds. Dom, the captain and pilot flying, started to initiate a go-around. As he started the maneuver, (poof!) the plane was out of the fog with the runway back in sight. The pilots continued the approach to a smooth and uneventful landing. On taxi, they reported to Tower, which was also Approach Control at that time of night, that a visual would probably not work for any aircraft behind them because of the fog. The controller replied that he had seen the cloud bank on final. But he never passed the information to the crew! (ASRS 427413)

Lessons when you can't see

It is beginning to be a recurring theme as we look at many of these cases. Hopefully, you can spot it by now: the role of complacency in aircraft accidents. Here we have another case of complacency, not on the part of the crew but on the part of the controller. Usually, complacency leads to a loss of situational awareness. However, in this case, the controller seems to have had his situational awareness in place. What he didn't have was concern. He saw the fog bank but didn't bother to report it to the crew. Whether this was because he didn't think it was important or because he was plain lazy is not clear. Whatever the reason, it should have been reported to a crew who was just issued a visual approach.

For the crew's part, there are a couple of lessons to be learned. This crew incorporated a solid practice of dialing in an instrument approach (preferably a precision approach if available) to back up the visual approach. This is an excellent habit. It serves as another reference for the visual approach and protects against an unexpected loss of visual references during the approach. When those visual references are lost, it is critical to immediately transition to instruments in order to maintain course guidance and spatial orientation.

The crew seemed to have every indication that a visual approach was appropriate. ATIS contained a good weather call; the crew could see the airport from 15 mi out (the darkness of the approach end precluded a view of the fog bank) and ATC offered a visual approach. All of these factors would have led you or I to accept the same approach. However, if there is any doubt as to whether visual conditions exist (there wasn't in this case), always request the instrument approach. That way, you can use whatever visual conditions do exist as a bonus.

The other pilot lesson is the consideration of the go-around. The captain began to initiate a go-around as soon as he lost visual contact with the runway. If visual conditions are recaptured quickly, it is probably okay to continue the approach. But be careful on this one. If you lose visual once, you may lose it again. There is no guarantee. So err toward the side of safety and go ahead with the missed approach unless you have good reason to do otherwise. Another trip around the pattern probably won't hurt, but another loss of visual on the approach may indeed hurt!

Where Did the Runway Go?

David, the captain of the Boeing 737 airliner, was cleared ILS Runway 4R circle Runway 19 at Newark International. The 737 was set up on base leg, near the end of the circling maneuver, at 1000 ft preparing to turn final, when Tower told him to extend downwind to 4 mi so the controller could launch a couple of airplanes off of Runway 4L. David turned downwind and proceeded to what he believed was 4 mi. He then turned base and then turned final to what he thought was Runway 29. At this point, David thought he had the VASI lights in sight and started his descent to landing. The Newark airport is surrounded by a tremendous number of lights (roads, warehouses, docks, and so on). It soon became apparent that he was not lined up on the runway, and about the same time, the tower asked the crew if it had the runway and VASIs, to which it replied "negative." "The crew was at 600 ft AGL. The controller suggested climbing back to 1000 ft and then pointed out the runway. The plane was approximately 3 mi from the approach end of Runway 29. The rest of the approach was normal. (ASRS 424156)

Helping out the controller

David, the captain, was justifiably steamed by this incident. He suggested, rightly, that when an aircraft is on a circling approach and is already circling and preparing to turn final, the tower should not ask that the aircraft make drastic changes to the approach to accommodate the tower. I agree. David adds that spacing needs should be handled earlier in the approach. He goes on to say that this is even more important at airports such as Newark that are very difficult to pick out with all the surrounding lights.

I include this study because who of us who have flown many instrument approaches hasn't been asked (ordered?) to help out the tower or approach with spacing needs by extending or shortening the approach? It has happened to nearly all of us. But some tricky things can occur during the accommodation. First, habit patterns are broken. Most of us aren't that proficient at circling approaches to begin with, so to ask an aircraft to extend during *base leg* of a circling approach is really asking a lot. I know that most of us want to help out; that is the kind of people we are. But we do have the option of saying "unable that request." Then the controller can deal with that, which may include another round of vectors for another approach. That option may not be desirable, but it is better than disorientation and CFIT.

David seems to have experienced some temporal distortion. The controller asked him to extend his downwind approximately 4 mi. Time (and distance) must have been proceeding faster than David thought because he flew well past 4 mi as evidenced by being so low when rolling out on final. It doesn't take long to fly 4 mi in a Boeing 737, even at circling speeds. But when our habit patterns are broken (by having a circling approach extended), time estimates can become confused. Using a DME readout (if available) would have been a good backup.

Where did the runway go?

The final and most important point to this case is the fact that the crew rolled out to the wrong runway—in this case a runway that didn't exist. The crew confused the lights because of the heavy concentration of lights in the area. Picking out the correct runway can be difficult enough in day VFR conditions. This was demonstrated by the airline crew that attempted to land on the wrong runway at Denver International in Chap. 4. Another example: I was once flying as a copilot when a crew began an approach to Lowry AFB in Denver. The only problem was the crew intended to fly to Buckley Air Base, which is a few miles away. The other problem is the runway at Lowry had been closed for at least a decade! The cars traveling on this "runway" kind of gave it away. Confusing the runway can happen even in day VFR conditions.

Choosing the correct runway can be even more difficult at night. I had a friend who was once shooting a nonprecision approach near minimums when he thought he spotted the approach lights. He began to slide over to the right in the direction of the lights (the localizer was nearly centered) only to find it was a local hamburger stand's flashing light on their new sign that had recently been erected near the airfield. A similar scenario caused the crash of a bomber aircraft in the 1980s.

There is a bit of a fine line to walk here. If the approach end of an airport is too dark, it can cause visual illusions and seem as though you are flying into a "black hole." On the other hand, if there are too many surrounding lights, as in Newark, it can be disorienting and difficult to pick out the runway lights. One of the crew's options in this instance was to try to figure some kind of distance off of a navaid on the field or via GPS. It then could have figured a 3-degree (e.g., 900 ft AGL at 3 mi) glide path to the runway and planned a departure from the circling altitude at the appropriate time. The crew may well have done this for the original circling approach and been unable to update that estimate because of time constraints on the extended downwind. If that was the case, it would have been best to decline the extension and ask for vectors for a new approach. Remember, it is nice to help out the controller, but not at the risk of your life or your passengers' lives. If the controller gives you an unreasonable request, politely decline it and go from there.

The Last Word on the Controller

In summary, we have learned several things about controller error. Number one is that controllers are human and will make mistakes. They go through training just like us, they have their good and their bad days just like us, but they are hopefully improving just like us. Controllers are susceptible to poor performance because of high workload and stress. They are susceptible to perceptual or mental sets; they may perceive something to be that just ain't so. They will commit errors, including errors of omission, when they simply forget to do something. Finally, they can be complacent like you or me and every once in a while downright lazy.

The pilot's best defense is to be as familiar with the area you are flying in as you can possibly be. In particular, know your minimum safe altitudes. Be on guard. If you hear nothing from the controller for a while, yet he or she seems to be communicating with everyone else, the controller could have forgotten about you. If you are not getting help from the controller, take evasive action if need be and clear it up later on the radio or on the ground. Finally, be willing to help out the controller—that's just professional courtesy. But don't let your help put you in a position of endangering you or your aircraft. If you can't help out the controller, say so, and he or she can then proceed from there. It's the best move for both of you. What changes will come in the future remains to be seen. What seems certain is that an air traffic control system that is currently straining under its workload will undergo change as air travel continues to grow.

We as pilots like it when it's "not our fault." But as we shall see in the next chapter, when it comes to controllers and instrument approaches, sometimes it is our fault!

References and For Further Study

Kern, T. 1997. *Redefining Airmanship*. New York: McGraw-Hill. Chap. 14.

Khatwa, R. and A. L. C. Roelen. April–May 1996. An analysis of controlled-flight-into-terrain (CFIT) accidents of commercial operators, 1988 through 1994. *Flight Safety Digest*, 1–45.

Reason, J. 1990. *Human Error*. Cambridge, England: Cambridge University Press.

Shappell, S. A. and D. A. Wiegmann. 2000. The human factors analysis and classification system—HFACS. DOT/FAA/AM-00/7.

9

Problems with Instrument Procedures

Two key inventions led to the advent of instrument flight. The first was the flight of Billy Mitchell in 1921. Mitchell was an Army officer who had boasted of the potential of this new idea called "airpower." He was convinced that it would make the difference in future armed conflict. His boasting led him to challenge the Navy to a little contest. Mitchell claimed that airpower could neutralize naval power, if not eclipse it totally. The Navy was willing to take up the challenge. The allies had captured a German warship, the *Ostfriesland*, during World War I. By treaty agreement they had to dispose of the vessel. The Navy considered this a chance to disprove Billy Mitchell and perhaps quiet his ranting. The authorities in Washington offered Mitchell the chance to test his claims against the *Ostfriesland*. Mitchell said he would sink her. On that fateful day in July of 1921, the naval brass gathered on a warship nearby in a festive party atmosphere. They were convinced they were going to witness "Mitchell's folly." Led by Mitchell, a group of airplanes dropped heavy bombs on and around the *Ostfriesland*.

Within a couple of hours the mighty ship listed and then slowly sank. Understandably, the naval party on the nearby ship came to an abrupt end. Mitchell was a hero— to some.

While this makes fascinating history, what does it have to do with our subject? The mission required Mitchell to be very resourceful. He had to not only develop heavy bombs (which he did), but he had to devise a way to strap them onto his aircraft and deploy them at the right time. More importantly, Mitchell was worried that bad weather would foil his flight on the appointed day. He had to be there that day, regardless of the weather. To this end, he develop an instrument that would help him fly. It was called the *artificial horizon*, and it was the precursor to the *attitude indicator*.

The invention of the attitude indicator came just a few years later. It was left to a man named Jimmy Doolittle and a company named Sperry. Doolittle would later go on to fame as the leader of the first bombing raid on Tokyo in 1942. Doolittle's Raiders took off from the *USS Hornet*, bombed Tokyo, and continued on to crash-landings in China, where they spent the remainder of the war.

Prior to this event Doolittle made another significant contribution to aviation history. In 1929, Doolittle set out to show that instrument flight was possible. Working with the Sperry company, he helped to develop what we now know as the moving horizon attitude indicator. This indicator is standard equipment on all aircraft in the Western world. Interestingly, the former Eastern bloc aircraft companies use the moving aircraft indicator, which may be a safer indicator. Such a discussion is beyond the scope of this book [for further information, see Previc and Ercoline (1999)]. Doolittle not only helped develop the instrument, but he made the first flight in instrument conditions on September 24, proving the concept of instrument flight was feasible.

We have come a long way since the 1920s. Aircraft are not only equipped with attitude indicators but with other instruments that make instrument flight possible. The altimeter, radar altimeter, vertical velocity indicator (VVI), airspeed indicator, approach plates, and navigation aids all contribute to instrument flight. Newer developments such as the Global Positioning System (GPS), the Inertial Navigation System (INS), precision approach radar (PAR), sophisticated autopilots and auto throttles, the Ground Proximity Warning System (GPWS), and the Traffic Alert and Collision Avoidance System (TCAS) have made instrument flying more doable, sophisticated, precise, complex, accurate, and, in some ways, more dangerous—all at the same time. Aircraft are able to go lower in more adverse weather conditions. As we will see throughout this chapter, problems with instrument procedures can lead to both CFTT and CFIT.

Surprise Approach

"I was the captain of the Learjet 35 that day in December. The ILS system at Palomar Airport had been down for the preceding two months, and the only available approach was a VOR approach. Because of winds, we planned to shoot the VOR and then commence a circling approach to land. It was my first officer's leg of the trip, so he was the pilot flying. The minimums on the chart were 1000 ft MSL for category B and C, the minimums being 672 ft above the ground. We set the approach minimums to 672 ft. During the first approach, we were forced to declare a missed approach, as we didn't see the runway. During the second approach, Palomar Approach called 'altimeter alert' to us, and it gave us a new altimeter setting. Once we adjusted the altimeters, we thought we gained some more space, and we descended another 100 ft to the minimums. Contact with the airport was made during this second approach, and we landed the aircraft safely. We taxied to the parking ramp and shut down the engines. As we were cleaning up and putting things away, we realized our error. I looked at the approach plate again. Instead of setting the approach minimums of 1000 MSL into our instruments, we had set 672 ft, which is how far the 1000 MSL minimums are above the ground. We broke the minimum approach altitude by 328 ft! To make matters worse, when Approach gave us the new altimeter setting, we actually descended another 100 ft to the minimums that we believed the new altimeter setting afforded us. This more than likely allowed us to make contact with the runway and facilitated our landing. I do not know why I would make such a fundamental error. I have made hundreds of approaches, and I am not weak in chart interpretation. We return to flight safety every six months for pilot recurrencies, and I receive excellent reviews. Neither crew member recognized the error, even though we failed to set the correct altitude twice." (ASRS 422969)

Déjà vu

This accident bears some eerie similarities to Eastern Flight 401 over the Everglades that we covered in Chap. 6. That was the accident where the crew became obsessed with the burned-out nosegear indicator while the altitude hold came off and the plane descended into the Everglades. The first similarity with that accident is the rather simple malfunction that occurred. In the Everglades crash it was a no-nosegear indication accompanied by equipment that didn't work as expected (the autopilot). In this case we have equipment that is not functioning (the ILS approach) and the crew simply setting the wrong approach minimums in the altimeter. Using our taxonomy of errors, this is known as an error of substitution. The crew substituted the AGL altitude for the MSL altitude. Obviously, observing correct approach altitudes is fundamental to good instrument flight. The remedy, of course, is attention to detail. The other remedy is good CRM, where each crew member backs up the other. In particular, the pilot not flying, the captain, bears a greater responsibility for not backing up the first officer at the controls.

The second similarity is the controller's vague questioning as to the real problem at hand. In the Everglades crash the controller asked, "How are things comin' along out there?" when he saw the aircraft 1100 ft low on his radar screen. Those were the days before controllers had automated altitude alerts. In this case the controller received an altitude alert, but instead of a direct query of the crew, he simply said "altitude alert" and gave it a new altimeter setting. This confused the crew as to the true intent of the controller's statement. The crew simply assumed it needed to enter the new altimeter setting because it was about 100 ft high, when in fact the crew was roughly 200 ft low at the time. The controller's vague attempt at help made matters worse. This is equivalent to a doctor treating the symptom and not the root cause of a disease. Clear and direct communication with standard terminology is vital to safety of flight. The Flight Safety Foundation's report on CFIT cites the lack of clear communication and terminology as a primary causal factor in CFIT.

Clear communication

It is easy to slip into sloppy habits of communication on the radio. Standard phraseology gets repetitive and sometimes we simply want to say things differently. The same tendency can happen with controllers. More than likely in this scenario, it was a controller who was just trying to slip the crew a clue without calling attention to its altitude deviation over the radios for the whole world to hear. In his effort to be friendly and protect the crew's ego, he might have caused its demise if circumstances had been slightly different. The lesson to take away: Be clear with your communication to controllers, and controllers need to be clear and direct with you.

"I Couldn't Let This Go Very Easily"

Jim and Kevin were flying the VOR DME Runway 25 approach to San Salvador in October 1998 at approximately 9:00 P.M. in moderate rain with a 1000-ft ceiling and about 6-mi visibility. When they broke out of the overcast at 600 ft MSL, the first officer, Kevin, commented that he felt that the plane was too low. They were at 3.8 DME on the 069-degree radial, descending to the 460-ft MDA. Jim disconnected the autopilot and climbed to 900 ft. He stayed visual at the base of the overcast until the VASIs became red over white and began descent again and landed uneventfully. They were a bit confused, so they examined the approach plate. They discussed it and didn't understand why there was a final approach fix (FAF) without an altitude. Upon returning home and researching the approach plate further, the company training department checked with the chart maker, and it was determined that the 1500 ft FAF was omitted from the plate and this plate's revision date was February 1998. At the final approach altitude, the Boeing 737 aircraft was at 600 ft MSL, not the required 1500 ft, a whopping 900 ft low. The plate was later revised and corrected, as the training department contacted the chart maker to inquire into the old plate. The crew is not sure if this plate previously had a FAF altitude or not, but it would like to think that you could convince it that there are no other plates like this in Central or South America. The reporter relayed the following: "When any aircraft—especially my Boeing 737 aircraft with me at the controls—is 900 ft too low at the FAF, I couldn't let this go very easily." (ASRS 430122)

Could it happen here?

This crew was very fortunate—fortunate that it broke out of the weather soon enough to realize how low it was because of an incorrect published approach. It was a good catch on the part of the copilot to sense that the plane was low. It was also good CRM in that he spoke up when he was uncomfortable. Perhaps the captain had set a tone in the cockpit that encouraged open communication. It's amazing that no one else had reported this error. The approach plate was put out in February 1998 and this was eight months later, in October. The crew took the proper error prevention measures after the fact. It analyzed the flight with a thorough debrief, and reflected upon the approach to determine how it ended up in such a predicament. A study of the approach plates revealed the answer. The crew then took the next step by contacting its company safety office to have the erroneous chart confirmed and changed so that future aircraft would be protected. If you don't have a company safety office, your best course of action would be to contact the Federal Aviation Administration (FAA).

The aircrew practiced good error mitigation in flight, when it arrested its decent and climbed back to 900 ft until it had the VASIs red over white indicating a 3-degree glideslope. As the saying goes, "White over white too much height; red over red you're dead."

Was there anything the crew could have done as far as error prevention is concerned? In other words, are there actions it could have taken prior to the incident to prevent such an occurrence in the first place? Such preventative steps are hard to imagine. The only possibility lies in recognizing that the FAF at 3.6 DME had no altitude depicted. This is generally standard procedure. Notice the crew was at 600 ft at 3.8 DME, prior to the FAF! Searching for irregularities on the approach plates would have been one option for discovering the error. The only other option would be to compare the MSA and to figure that you should not be too far below that until inside the FAF. Either of these suggestions would be easy to overlook by the vast majority of us, me included. But it is something to be mindful of. We come to at least expect that our approach plates are accurate and complete, and that we can trust them. But this was Central America. Surely this couldn't occur in the United States, could it?

The Lone Star State

"I was over the cotton fields of the west plains of Texas, bound for Lubbock, home of Buddy Holly and Mac Davis. I was in my Cessna Caravan being vectored to intercept the localizer backcourse for Lubbock International (LBB) outside FRITS intersection (the IAP) at 5200 ft. Around 10 ILBB DME, I started my descent from 5200 ft to 4800 ft as published. At approximately 6 or 5 ILBB DME, the controller (I believe I was now on the tower) notified me he was receiving a low-altitude alert. I checked my altitude and it read 4800 ft, and I started a climb back to 5200 ft. Now I'm passing FRIER and I'm descending to 3820 ft. I had ground contact but did not have the runway environ-

ment in sight. I was two dots right of course and correcting. The 'rabbits' (approach lighting) came in visually at
ENZIE, and I landed without further incident. After taxiing
to the ramp I asked the tower what altitude they thought
I should be at between FRITS and FRIER, and I was
informed that I needed to be at 5200 ft. The approach
plate clearly showed 4800 ft as the correct altitude. Having
to recheck my approach plate at the final approach fix
made for extra workload at a critical time of the approach.
I climbed back to 5200 ft, because I would rather be alive
high 'than be dead right.' I turned this information over to
our safety director and suggested he check with commercial to see if there was a mistake on the approach plate. I
checked with other company pilots, and they had also
gotten a low-altitude alert when making that same
approach down to the chart depicted 4800-ft altitude.
Therefore, when the controller stated that I should be at
5200 ft, it seemed reasonable because of the altitude
alerts. There are two high transmitting antennas just off to
the left of the final approach course, which probably
caused the altitude alerts. I also checked the government
chart and found it to show the 4800-ft altitude." (ASRS
430947)

Good technique and philosophy—it can happen in the United States

In our previous case we saw a missing altitude from the
approach plate at San Salvador. In this case, the altitude
published is questionable. The charts consistently indicate 4800 ft, but the approach controller relayed 5200 ft,
which did take care of the low-altitude alert approach
was receiving on the Caravan. Again, the pilot did a nice
job of reflecting on what occurred in order to prevent a
reoccurrence. This is a key to improving aviation skills:
reflecting on lessons learned. He followed up on the

problem by reporting the incident to the company safety office, as did our last pilot.

There is another factor to consider in this situation: the accuracy of the pilot's approach. The pilot states that he was two dots right of course and correcting. Since this was a backcourse localizer, that means that the pilot was actually left of course. A backcourse localizer, or any backcourse approach, is counterintuitive in that you turn away from the course indicator in order to correct to course. A good backcourse approach is a feat of mental gymnastics. By being left of course by two dots, the pilot placed his aircraft over the two high transmitting antennas, setting off the altitude alert. Had the pilot been on course, it is likely that no alert would have been registered.

So how far off course is too far? This pilot was two dots off and got into trouble. A good rule of thumb is that if you get more than two dots off course, it is time to go missed approach. If you remain at or less than two dots, you are probably safe to proceed. Inside the final approach fix or once on glideslope, you should start to get concerned if you deviate more than a dot off course, especially if weather is near minimums. This pilot was right in the region to seriously consider a go-around.

The pilot did make one good decision, and he left us with a good quote. When the controller told him 5200 when he showed 4800, he didn't argue but immediately started climbing. He realized there was little benefit in debating the merits of the published approach. As he said, "I would rather be alive high than be dead right." When it comes to low terrain, there is no time to argue. The pilot exhibited good technique and philosophy.

It seems from this case in Lubbock that there can be problems with the published approach in the United States as well as in Central America. And as our next two cases will show, the problem is not limited to the Lone Star state.

Overboost That Mountain

Glenn and Phil were at the controls of the Super MD-80 heading into the Bay Area. San Francisco International, Navy Alameda, and Oakland International are all in close quarters. When they were on the Locke 1 arrival to Oakland in IFR conditions descending to 5000 ft MSL, GPWS announced 'Terrain, Terrain.' Phil immediately initiated an escape maneuver and a climb back to 6000 ft. During the escape maneuver, both engines reached a peak EPR (exhaust pressure ratio) of 2.2, which is an overboost of the engines. The terrain avoidance maneuver and subsequent landing were uneventful. After landing, Phil placed a phone call to Bay Approach in regard to the incident. Approach advised there was no loss of separation. It also said that GPWS events do occur now and then on this arrival due to the Bay's airspace restrictions that require aircraft to overfly a local peak, Mt. Diablo, which rises above 3800 ft. Following the flight, maintenance examined the overboost of the engines to 2.2 EPR. Mainten-ance was forced to remove the aircraft from service, and then extended flight was canceled until further engine inspections could occur. Oakland personnel say that the company had three of these escape maneuvers recently and the jumpseating pilot on board said it happened to him also. The captain was justifiably upset. He considered this an ATC-induced escape maneuver that occurs too often and suggested that a minimum altitude of 6000 ft be in effect until aircraft pass Mt. Diablo. (ASRS 428677)

Overboost, should we?

This type of incident would not have occurred 25 years ago. Aircraft were not equipped with GPWS; therefore, no alarm would have sounded on the aircraft and the crew would have proceeded merrily on its way. Advances in technology bring their benefits and their baggage. Things have changed since the advent of GPWS. Crews do react to GPWS. As reported in Chap. 6, initially GPWS warnings were often overlooked by aircrews because of the high rate of false alarms. "The boy who cried wolf" syndrome led aircrews to disregard alarms on a frequent basis. This case study illustrates how far we have come from that position.

This crew reacted quickly and aggressively to the alarm. The captain immediately initiated the escape maneuver, overboosting the engines in the process. Is this a problem? It is only a problem for the maintenance folks. The crew did the right thing. When an alarm goes off, the crew needs to take immediate evasive action. That is a good lesson for you and me. The GPWS alarm is one of the few stimuli that should cause an immediate reaction from the crew without it first thoroughly analyzing the situation. A TCAS alarm is another instance where the crew should react quickly while trying to spot the traffic visually. Most of the time we should "wind the clock" and take our time when something out of the ordinary occurs on the flight deck. A GPWS alert is not one of those times.

Even though the crew did the right thing with the quick response, it is justified in its anger concerning the false alarm. The arrival should be changed. This is just another example where the approach plates are not infallible. Listen as another captain describes this phenomenon.

Another Imperfect Approach

"While flying the approach to the localizer backcourse Runway 28L at Boise, we were being vectored to the final approach course in our Boeing 737 while descending to the assigned 6800-ft altitude. Our path over the ground took us to the 20 DME radar fix, where there is 5360-ft terrain. Our descent rate (approximately 1000 fpm) and the lateral closure rate triggered the GPWS 'Terrain' warning. I executed an immediate missed approach, intercepted the localizer, and climbed to 8000 ft MSL, while my copilot requested vectors for another approach, which was conducted without incident." (ASRS 429442)

The precision of the GPWS

Again, we see a case of a GPWS warning and evasive maneuver where the crew would have been none the wiser 25 years ago. It may be as time progresses that approaches will be changed to accommodate the sensitivity of the GPWS. The captain recommended in this case that the vector altitude be changed to 7500 ft, or even 8000 ft MSL for the 20-DME-to-15-DME transition segment. He further asks that they provide a cautionary note regarding a possible GPWS warning if altitudes are not changed. A third possibility is to revise controller procedures to vector aircraft inside of the 20 DME fix.

Advances in GPWS will also allow them to become more precise, only sounding when an actual vector toward the ground is likely. GPWS has come a long way, but it is not perfected yet. In the meantime, continue to follow the example of both this Boise crew and the crew going into Oakland. When the GPWS alarm sounds, initiate the escape maneuver and investigate later.

Build Your Own Approach

It was a warm summer evening in the windy city of Chicago. The Cubs were completing a baseball home-stand at Wrigley as Sammy Sosa chased Roger Maris and Mark McGwire in the home run race. The first officer and pilot flying, Cole, briefed the NDB Runway 14R approach at Chicago O'Hare (ORD). Cameron, the experienced, 12,000-h captain, was sitting at his side. This was the crew's first pairing and second leg of the day. Nightfall was coming on as the crew descended in VFR conditions and picked up ground contact from about 7000 ft AGL. An NDB was a rare occurrence in the crew's operation, so it decided to "build" an NDB approach with the FMC. This little trick would give the crew a backup on its position and a way to anticipate the crossing of the final approach fix. The crew overlooked only one thing. Its constructed approach depicted the "OR" beacon (a step-down fix) 5.2 mi north of the beacon's actual position.

It was a Monday evening, and ORD Approach Control was very busy, which is often the case in the early evening. The crew was given several radar vectors; the final one left it on a 140-degree heading, but the controller did not clear the crew for the approach. A strong westerly wind was drifting the aircraft east of the final approach course. Finally, it received approach clearance and Cole had to scramble to reintercept the course and get the air-craft in the landing configuration prior to the final approach fix. A hurried turnback was necessary to inter-cept the course, because the FMC showed the crew almost at the NDB. Raw data were being displayed on both the RDMI (radio distance magnetic indicator) and the Nav dis-play. Cameron and Cole thought they were very close to the beacon, and when they saw a 10-degree swing on the ADF needle, they assumed they were at 'OR.' (You know what happens when you assume.) They both saw

the needle start to swing, and Cole called for the next altitude. Cameron selected the next altitude, started the time,
and got ready to call the tower. Tower frequency was
busy, so he didn't get a call in right away, which again is
not unusual at ORD. They had descended about 500 ft
when approach control said they had a low-altitude alert
on the aircraft. They climbed back to 2400 ft, and then
they realized what had happened. They could see that
there were no obstacles in their vicinity, and they were
still about 1200 ft above the ground. Then they both
looked at the ADF needle and saw that it was still pointing up. They felt really stupid. The FMC showed the NDB
further out on the approach than it should have been
because of their error. The NDBs and other nonprecision
approaches they fly at recurrent training require a descent
started almost immediately at station passage. Otherwise,
the aircraft won't be at MDA in time for a normal approach
to the runway. (ASRS 414116)

Filled with lessons

In all of the ASRS reports I have reviewed, none are filled
with more valuable lessons than Cameron and Cole's
flight into Chicago O'Hare. It is difficult to know where to
begin. First, they were shooting a rare approach, a nonroutine operation. Of all the nonprecision approaches
flown today, perhaps none is as nonprecise as the NDB.
An NDB is simply not that accurate. The crew does not
state why it chose this approach, but I personally would-
n't go out of my way to choose it in actual IMC. Save these
types of approaches for training during day VMC. Second,
the crew was wise to choose a backup approach for the
NDB. It was planning ahead, which is always a good idea.
However, it seems that no other approaches must have
been available, as the crew "made its own" by the use of
the FMC.

The proper programming of the FMC is essential. Perhaps nowhere else on the aircraft does the old saying fit as well as with the FMC: "garbage in, garbage out." Most readers will remember Korean Airlines Flight 007 that was shot down over the eastern portion of the then Soviet Union in the 1980s. The crew had departed Alaska and was heading for the Far East. As best we can tell, it punched a wrong waypoint into the FMC and wound up way off course—so off course they ended up in Soviet airspace. Soviet fighters were scrambled to intercept KAL 007, as radar suspected it was a spy plane. The fighter pilot visually identified the aircraft and was ordered to shoot it down—an order he complied with.

Cameron and Cole made the same error here, to a much smaller degree. They punched in the coordinates for the "OR" beacon, a step-down fix on the approach, but they missed the coordinates by over 5 mi to the north. This caused them to start their descent 5 mi early. The remedy: Watch what buttons you push and then double-check them. The crew accomplished no double-check of a backup to a rare approach. Its poor CRM almost cost it.

One possible reason for the crew's poor CRM could have been the fact that Cameron and Cole had not flown together before this day. This was only their second flight together. When crew members are unfamiliar with one another, it takes a while to establish good teamwork. It is similar to what we see at an NBA All-Star game. The players obviously have lots of talent, but they aren't used to working together, and you see lots of coordination (e.g., passing) errors. Often a less talented team that has played together often can give the All-Stars a run for their money, as we saw at the 2000 Sydney Summer Olympics. The USA basketball team won a close victory over a team with much less talent.

Primary backup inversion

Remember, the FMC was the backup for the approach. The crew made a second error by descending without proper deflection of the ADF needle. The needle starting to drop (roughly 10 degrees) is not a sign of station passage. You should wait until the needle is clearly past the 3 or 9 o'clock position on its drop. The crew simply demonstrated poor instrument flying procedures. Again, a double-check would have caught the error.

By flying off of the FMC and using the ADF needle as a backup, this crew demonstrated a concept known as *primary backup inversion.* This is where the primary system gives way to the backup system without good reason. The reason usually is that the backup is simpler and requires less effort from the pilot. An example of this concept is when a crew stops visually scanning outside the aircraft for traffic and relies exclusively on the TCAS to do its clearing. The TCAS is designed as the backup for the eyeballs, not vice versa.

Primary backup inversion caused the crash of a domestic airline jet a few years back. Airline jets need the flaps down to take off, as do most aircraft. Part of the pretakeoff check is to ensure that the flaps are set at takeoff position. The crew can do this visually by looking back out the cockpit window and by manually checking the flap handle in the takeoff (usually 20-degree) position. It seems that another practice had pervaded the culture of the airline. If the throttle is pushed beyond the 80 percent position (as it would be for takeoff) and the flaps are not down, a warning horn sounds. Well, instead of checking the flaps visually as per technical order, the crews began to do the "lazy" check by simply running the throttle up quickly past the 80 percent position and then back to idle. If the horn did not sound, the crew could be assured that the flaps were in takeoff position...well, sort of. The airline crew in question ran the throttle up and then back down and got no horn. It was then cleared for takeoff, a takeoff it attempted to accomplish without the flaps down. The horn had malfunctioned; the crew didn't have the lift it needed and crashed after liftoff. The primary system is what we should deal with; if it fails, we have the backup.

Everything pressed together

This crew had several factors converging on it at once. The approach and the tower were very busy with traffic, so there was little break on the radios. This high traffic requires intense listening for a break to speak and for when your aircraft is called out among all the others. The crew also experienced time compression, as it was being pushed by winds away from the course that it was waiting to be cleared for. The poor vectors caused the crew to be rushed to complete all required checks and configurations. Because it thought the step-down fix was much closer than it actually was, it felt even more rushed. It was probably overloaded with information as the fix grew closer on nonroutine NDB approach with lots of drift. It was watching both the FMC and the ADF needle and it had bad information coming from the FMC. This portion of the approach has high workload when all the factors are in your favor. But this crew had things stacked against it.

Bad transfer

The crew was set up for an improper descent out of the fix by its training practices. This is known as *negative transfer of training*. All of the NDB approaches the crew practiced in the simulator required it to depart the altitude immediately after station passage. If the crew didn't, it wouldn't get down to the MDA in order to visually acquire the runway in time to land the aircraft. This was not the case that evening in Chicago. The crew had a certain mental set that this had to be done. We discussed mental sets in detail in Chap. 4. It is hard to break out of a mental set. The crew was primed for a descent at the first sign of needle deflection. Ordinarily, an expeditious descent at the FAF is a good practice, but in this situation it contributed to the crew's feeling of time compression. This is analogous to parking your car in a certain part of the parking lot on a routine basis. One day you park your car on the other side of the lot for some reason. Chances are when you come out of your office, you will head to the regular side of the lot to get your car.

Negative transfer of training can be deadly. I remember hearing an accident report of a police officer who was gunned down in a shoot-out with felons. When they found the officer, he was dead, with five bullets neatly lined up next to him. That is the exact layout the police practiced when at the shooting range. Police were required to line their bullets that way on the range in order to help account for ammunition and to know when to have their targets scored. There is no requirement to line up your bullets in a shoot-out for your life. Similarly, there was no requirement for this crew to get down right at station passage to successfully execute the approach. Examine your training practices. Do they set you up for failure in the operational world?

The next three examples illustrate common day-to-day errors pilots can make to get themselves into trouble. We began this chapter with a crew who dialed the wrong circling approach altitude into its altimeters. This is a elementary mistake with potentially grave consequences. I will conclude the chapter with several short examples of similar simple or common errors that were almost costly.

Watch Those Towers

"We were a light transport on an air taxi operation heading into Joplin (Missouri) Regional. Weather conditions were mixed VMC and IMC. Once we were cleared for the localizer/backcourse approach and established, we misidentified a step-down fix and began our descent to minimum descent altitude (MDA) approximately 4 mi early. In our descent, we noticed a tower in our path, and we deviated around the tower and concluded the approach. What the crew learned from this is to carefully identify all step-down fixes and thoroughly brief the approach." (ASRS 418944)

This crew committed a simple error in misidentifying a step-down fix and started down toward the MDA. This is the same error committed by the Learjet on the NDB approach into Palomar at the beginning of the chapter. The step-down fix was there for a reason—to give proper clearance over a tower. Several of the case studies in this chapter on poor instrument procedures consist of more difficult approaches. I don't think that is an accident. The Flight Safety Foundation found that 60 percent of the CFIT accidents it studied that occurred during the approach phase occurred with nonprecision approaches. Furthermore, difficult approaches such as an NDB or a backcourse localizer demand more of the pilot and are less precise than other published nonprecision approaches. By his own admission, this pilot had not given the approach its due respect. Any approach, but especially the ones we are discussing, requires a thorough study and briefing among crew members. This is part of proper preparation before flight and proper procedure in flight. This crew was remiss in its duty. Had it thoroughly studied and briefed, it likely would not have missed this step-down fix.

The crew was fortunate that it had maintained a good visual scan outside of the aircraft. It was able to visually acquire the tower and maneuver to avoid it. In many of the cases we have studied in this chapter, the control tower has alerted the crew as to an important altitude deviation. Furthermore, the crews were often warned by the GPWS of a hazard. This crew was not fortunate enough to have either of these backups in this situation. You may not be fortunate enough, either. The lesson is, you can't depend on ATC or automation to catch your errors. You are primarily responsible.

Coffee and a Sandwich

"On January 18, 1987, I had been a copilot on the C-130 Hercules for about a year. That was my first assignment out of pilot training. Being stationed at McChord AFB in Tacoma, Washington, we flew many scheduled missions into Alaska hauling cargo and performing airdrops of people and equipment. I had spent a fair amount of time in many of the urban and rural airfields of Alaska and felt comfortable flying into some of the most remote and challenging airfields in the world. On January 18, our crew was participating in an exercise called Brimfrost '87 that had our unit operating 24 h per day moving people and equipment all over Alaska. My mission was to fly a round-robin from Elmendorf to Sitka to Allen AAF back to Elmendorf. The aircraft commander who I was scheduled to fly with during the exercise was a 'crusty ole major' who had more time in the C-130 than I had alive. He was one of those officers that had more interest flying the plane than in flying the desk, so consequently his gold leaves were never going to turn silver. He seemed happy with that and was due to retire to a quaint little cabin with his wife in the hills near Mt. Rainier very, very soon. Flying with this guy was going to be a little slice of heaven.

"The first leg to Sitka, Alaska, was beautiful. Scattered clouds hid bits and pieces of the many fjords that jutted to Alaska's interior. The weather in Sitka was going to be scattered layers with an overcast deck on top. The approach in use was the LDA/DME 11. There was a TACAN arcing maneuver we used to get lined up with the final approach course. The TACAN is located on a mountain next to Sitka off to the right if flying the localizer approach. We were on top of the overcast and could see the snow-capped mountains above the clouds past the airfield. The major gave a cursory approach briefing, and I figured this was routine for him. He must have been here dozens of times, although it was my first. There was light conversation going on between the engineer and the navigator. The mood was relaxed—too relaxed.

"The major arced nicely toward the final approach course, computed and turned at his lead radial, and rolled out beautifully on the inbound course. We started to let down into the overcast when the navigator, in a confused voice, chimed in that he had some big black thing on the radar screen that we were heading right for, and he asked if we were aware of it. At first, we brushed him off with a 'Yeah, there is a mountain to the right of the course.' We then took a closer look at the approach plate and realized at the same time as breaking out from the clouds that we had made a huge mistake. Sure enough, the radar was correct. We were about 5 mi from the mountain that the TACAN was located on for the arc to final, but the final approach course was off of the LDA/DME located at the field. We were flying the final approach course off of the TACAN! We never noticed that we needed two navaids for the approach. If the clouds were thicker and we never broke out or if the navigator hadn't been backing us up on radar, we'd be dead. We slid over to the left with the runway in sight and landed uneventfully. I was shaking, thinking to myself, 'So this is operational flying. Wow, that was close!' I was looking forward to a good chat with the crew to figure out how we had gotten ourselves into this mess. I looked to the seasoned veteran to my left. 'Let's go grab some coffee and a sandwich,' he bellowed. Coffee and a sandwich? Didn't he want to figure out what had happened? 'Nah, let's not talk about that. I'm hungry.'

"I learned that just because I'm not in charge, I have just as much responsibility for the safe operation of the flight as the aircraft commander. Rely on but don't blindly trust your crew members. Brief thoroughly. Fight complacency. Stay sharp on procedures. Mountainous terrain *always* requires your full attention. Pay attention to details. Ask questions. Don't assume (it was the major's first time here too). The insignificant things will kill you. Pray daily."

A simple error

This is another simple error we could all make. The crew is doing a published penetration with an arc to intercept the inbound radial for the approach. The approach requires a change of navigation aids to fly the final portion of the approach. Determining the lead radial is key. Whether the approach plate indicates a lead radial for the turn and switch of instruments or the crew figures one for itself based on that aircraft's turn radius, a good lead radial can help ensure a smooth and successful approach. Of course, the lead radial is only helpful if we use it.

How could four people on one aircraft miss this error? There are probably several reasons, all of which go back to CRM. You have probably the two most dangerous folks you can have at the controls. In one seat you have a very experienced aircraft commander at the end of his career, with nothing to prove or shoot for, other than 20 years. In the other seat you have a brand-new, eager copilot wanting to prove himself as a valuable new member of his organization. Those are very fine qualities that can help to overcome important qualities he is lacking—namely, skill, experience, and situational awareness. These come only with time. I've heard many experienced air commanders say that the two folks they worry most about in the cockpit are the new guy in his first few months and the really experienced guy. The experienced guy is likely to fall prey to complacency, and the new guy doesn't know enough to know when he is in trouble.

What about the other two crew members? The FE said nothing as he covered his aircraft gages. The really good FEs in the military learn to read approach plates and conscientiously back up the pilots on the approach. The average FE is content to just cover his or her territory. This strikes me as an average FE. The aircraft commander should encourage the FE to become an excellent FE. The navigator was trying to warn the crew of danger. He was doing his job in backing up the approach. However, his warning fell on deaf ears at first. The new copilot was partly not concerned about the nav because he thought he knew what was going on and the experienced aircraft commander obviously wasn't worried. The experienced aircraft commander was probably just complacent. Anytime a valued crew member asks a question of the aircraft's status, we owe more to that crew member than a "Yeah, there's a mountain out there." We should take a step back and ask ourselves, "Is there a problem here?" Had the crew done that in this situation, it would likely have caught its error with the navaids sooner.

The most disturbing part of this story is the aircraft commander's response on the ground. Grabbing a coffee and a sandwich is fine. But you need to thoroughly debrief the flight as well. Debrief over the coffee and sandwich if you like, but debrief the flight. The aircraft commander is setting a poor example for a young copilot in this situation. As his mentor he is essentially saying, once the flight is over, forget about it. That's not a lesson experienced folks should be teaching; that is an unprofessional attitude. It is a failure of leadership and responsibility. A thorough postflight analysis is a vital part of error prevention.

What Page Was That On?

Ron and Sam were in a business jet on an approach into Medford, Oregon. The local area is hilly and mountainous. Because of the high terrain, the approach is a stair-step-down. It was mixed VFR and IFR conditions. Something didn't seem right to Ron. He checked his approach plate, and Sam checked his. Neither of them had the right approach plate out! They had a Medford approach out, but to a different runway. They immediately went missed approach. In the weather, it would have gotten them! (personal communication)

No need to go on and on with an analysis of this incident. Obviously, the pilots aren't going to fly the approach very well or very safely if both have opened the approach to the wrong runway. Attention to detail solves this problem. They were wise in not trying to continue the approach. The right decision was to go missed approach ASAP.

Where Does This Leave Us?

It should be clear that we have come a long way since the days of Billy Mitchell and Jimmy Doolittle. The road is fraught with obstacles along the way. The advent of instrument flying has opened the vast majority of the world up to the feasibility of regular flight. However, with this increased freedom comes responsibility. Responsibility and freedom always go hand in hand.

The last couple of case studies show how the little things make a big difference. Getting out the right approach plate, setting in the correct navaids, correctly identifying all fixes, especially step-down fixes, are all basics of instrument flight. Basics in instrument flight are essential. In fact, the fundamental basics are almost always important. Vince Lombardi, the great Green Bay Packer football coach, became upset with his world champions after one particularly poor game. At the next practice, he held a football up into the air and said something like "Gentleman, today we go back to the basics. This is a football." Don't forget the basics.

We also saw that some approaches are more difficult than others. All approaches deserve thorough and complete study in advance and a solid briefing to refresh all the crew members in the air. NDB and back-course approaches in particular bear extra care, as they are nonroutine to many of us. Remember as well that published approaches can be in error. A thorough study of the approach may reveal the error. A check of the NOTAMs is also essential. However, if an approach is in error, be prepared to take evasive action at the first sign of trouble indicated by the controller, the GPWS, or the other crew member. Back each other up!

Is there anything else that can be done to prevent CFIT and CFTT from poor instrument procedures? Research conveys a couple of suggestions. The first is to stay proficient. This isn't rocket science, but sometimes we let it slip. Make sure you get a good workout when you go "under the hood" for IFR practice with another pilot. I know of pilots who used to "sneak a peek" at the runway when flying a hooded instrument approach to help align themselves with the runway. They would tell me, "One peek is worth a thousand crosschecks." That may be so, but you don't always get one peek when the weather is at minimums. Don't cheat yourself by looking good during day VMC and looking bad when the chips are on the line and the weather is at minimums and you can't get the airplane down.

Two researchers at the Douglas Aircraft Company, Richard Gabriel and Alan Burrows, showed way back in 1968 that pilots can be trained to improve visual scanning in order to give the pilot more time to think about factors outside and inside the cockpit. An efficient scan can give you more time to get the big picture on the approach straight in your mind. It will also allow for a greater ability to identify threats outside of the aircraft (e.g., other aircraft, towers) that threaten to impact your future. A number of publications are available on how to improve your scanning. You can also pull aside a respected and experienced aviator and pick his or her brain on the subject. However, this must come from a motivation to improve on your part. It also concerns planning properly before flight, a subject that we will examine next.

References and For Further Study

Adams, M. J., Y. J. Tenney, and R. W. Pew. 1995. Situation awareness and the cognitive management of complex systems. *Human Factors,* 37(1):85–104.

Bellenkes, A. H., C. D. Wickens, and A. F. Kramer. 1997. Visual scanning and pilot expertise: The role of attentional flexibility and mental model development. *Aviation, Space, and Environmental Medicine,* 68(7):569–579.

Gabriel, R. F. and A. A. Burrows. 1968. Improving time-sharing performance of pilots through training. *Human Factors,* 10:33–40.

Gopher, D., M. Weil, and T. Bareket. 1994. Transfer of skill from a computer game trainer to flight. *Human Factors,* 36(4):387–405.

Previc, F. H. and W. R. Ercoline. 1999. The "outside-in" attitude display concept revisited. *The International Journal of Aviation Psychology,* 9(4):377–401.

10

Planning and Time

Proper planning prevents poor performance. It seems that a lack of planning is often associated with CFIT and CFTT. Closely related to the concept of planning is time, the fourth dimension. Temporal distortion plays into CFIT and CFTT. Time and planning often go hand in hand as we read the ASRS reports. It seems like these two concepts are inextricably linked. Time pressure and planning, planning and rushing—how often have you been there? How can we hold back time?

What Goes Down Must Go Up

"We were descending from cruise altitude to land at the Chippewa County (Minnesota) International Airport (CIU), approaching the airport from the southeast. We were in a regional commuter that Sunday morning. We were expecting to get a visual approach to Runway 33 and were anticipating that the controller would give us a lower altitude than the assigned 4000 ft MSL (which we were flying at). At 4000 ft we were just in the base of the overcast layer and had intermittent ground contact below. This is the first time either I or the copilot had been to CIU. We asked the controller for lower, as we wanted to break out and go visually. When he advised us that this altitude was as low as he could give us, we were about 15 mi from the airport. I realized we wouldn't get in visually, so I asked for a vector to the nondirectional beacon (NDB) approach to the landing runway. We were practically on the inbound course anyway. We were rushed and didn't look at the approach plate closely enough. There is a step-down fix on the approach at around the 30 DME fix from another VOR's crossing radial. At this point we could descend to 2400 ft to VFR conditions, which is what we wanted to do (just use the approach to get below the base and go visually). When we tuned in the crossing VOR's radial and DME to determine our step-down fix, it read 28.5 DME. We proceeded down to 2400 ft, as we were on the inbound course established and believed we had crossed the step-down fix. We broke out visually and had the field in sight. The controller told us to climb back to 4000 ft, as we had descended too early. We climbed back into the clouds and figured out that instead of the DME normally counting down as we expected, it should have gone up based on the angle of the crossing radial. At 28.5 DME, we were actually 1.5 mi away from the step-down. We thought we were 1.5 mi inside (past) the fix.

"Some factors that led us into this situation were that we were expecting a visual approach initially and hadn't expected to use the approach until we were very close to the airport. This led us to rush the approach brief and not carefully review the approach plate as we would normally do. This could have been a deadly mistake in mountainous terrain. Next time I will try to establish from the controller what approach to expect further out from the airport. If the weather is marginal, I will ask for the approach further out and have more time to set up and brief the approach." (ASRS 415915)

We assumed

The case study provides a nice transition from the last chapter, as the crew exhibited poor instrument procedures. However, there is a deeper underlying problem here. It was the crew's first time into this airport, so it was heading into unfamiliar terrain. It didn't know what to expect at Chippewa, but it did have some expectations. It had a mental set that upon arrival it would get a visual approach. The crew may have been led to this expectation by a weather forecast; it is hard to tell. What is clear is that the crew assumed it would get a visual approach and it was not ready for an IFR approach. Everyone knows what happens when you assume.

This crew also exhibited a failure to plan. It should have briefed an instrument approach during the mission-planning phase of the day. This is a good idea, even if VMC is forecasted at your destination. Things can and will often change. Even as it drew closer to the airport and it was evident that conditions were questionable the crew still assumed the approach would be visual. The crew should have spent this time planning an instrument approach. It took the controller's information that a lower altitude was not possible to shake the crew from its poor assumptions.

Sometimes it goes up

By the time the crew awoke from its fantasy, it was behind the eight ball timewise. The crew's lack of planning led to time compression and a rushed approach. It did not have time to thoroughly study or brief the approach. Because the crew did not study the approach well, it came to another faulty assumption: the VOR would count down to its fix. Usually a crew is using a VOR on or near the airfield of intended landing. In that case, the mileage usually counts down. But this crew was using the VOR from another location off of the airfield; sometimes the mileage counts up. As the crew pointed out, this could be a deadly mistake in mountainous terrain. Fortunately, it received some good backup from a watchful controller.

Because of a lack of planning, this crew experienced temporal distortion. The adrenaline release caused things to be rushed and confused in the crew members' minds. The crew prematurely jumped to a conclusion—namely, that the VOR mileage would count up. All of this would have been avoided through proper planning. Our next case study illustrates another hazard associated with a lack of planning.

Bummed Out

Dale had some important business to take care of in Des Moines, and he was eager to get there. It was a snowy Wednesday. Soon it became clear that Dale's destination, Des Moines International, was becoming impossible because of snow, so he requested vectors for his Cessna 210 to divert to his alternate, which was near minimums (unforecasted). The alternate shouldn't have been a problem, as Dale is instrument-rated and proficient. He was very "bummed out." He had an important client to meet with, hoping to tie up some loose ends on a business deal before the end of the quarter. That didn't seem possible now. Arriving at his alternate, he was vectored onto the localizer. Approach asked him to keep his best forward speed to the marker. He replied with a "Roger, wilco." For some reason, he misread the glideslope indicator and thought he was inside the outer marker (OM) when he was still outside. He started to descend. He was 300 ft below glideslope at the marker when ATC called "Low altitude alert." Dale snapped out of it. He corrected up to glideslope and landed uneventfully. (ASRS 424505)

Check your bags at the door

Dale was motivated to get to Des Moines. There was a deal to cut. He was severely disappointed to find that a freak snowstorm was going to keep him from landing there. His disappointment was acute; this was an important business deal. This disappointment translated into what is known as emotional baggage or emotional jet-lag. It occurs when there is disappointment. You can't get that disappointment out of your mind, and it distracts you from the task at hand.

He took this emotional baggage with him to the alternate. He hadn't given much thought to the alternate because Des Moines was forecasted to have pretty good weather. However, he was wise in having an alternate picked out. Before any flight, it is just good practice to identify alternate fields in case they should be needed. These fields should be along your route in case of emergency and near your destination in case of bad weather. It is good to even circle these on your chart or have them handy on a piece of paper or grease board so you can quickly switch gears if need be.

Forgetful in my old age

The reason it is so important to have the alternates written down is that it saves the short-term memory from having to keep up with such details. This frees up short-term memory to be able to focus on the chores of the routine flight such as radio frequencies, navigation, and basic aircraft control. We know that short-term (or working) memory has a capacity of seven plus or minus two "chunks" of information. This is best illustrated by a telephone number. The number 261-4173 contains seven chunks of information. However, if you split up the information into 261 and 4173, you now have two chunks of information. That is why business dealers often have catchy numbers such as FIX-AUTO for a mechanic. The phone number FIX-AUTO is essentially one chunk of information. The lesson is that freeing up short-term memory decreases workload for the pilot. This was confirmed by Berg and Sheridan (1984), who studied pilot workload and found that simulated flights heavy with *short*-term memory tasks had higher subjective and objective workload ratings than did flights with high *long*-term memory demands. So write things down, and they won't burden short-term memory or cause you to scramble to find an alternate when the need arises.

Keep your speed up

The fact that Dale carried emotional baggage to his alternate was worsened by the controller's request that he keep his speed up to the marker. In combination, these items overloaded Dale. He was not mentally prepared to fly an instrument approach, which requires concentration and mental self-discipline. He was in a figurative fog. He was lucky that the controller snapped him out of it. Once he discovered he was 300 ft below glideslope, he would have been wise to go missed approach. At 300 ft below the glideslope, he was likely more than two dots below on his indicator. If you are that low on glideslope, it is time to go around. Besides the obvious safety concerns with CFIT, going around would have allowed him to refocus and mentally prepare a second time for the approach.

The Great Outdoors

It was flying as it was meant to be. Jordan planned the takeoff in the Stearman with a passenger and a full load of fuel from an uncontrolled 2500-ft grass strip. The windsock indicated calm winds; however, the field was surrounded by trees, which can mask or cause swirling winds. Run-up and takeoff were normal with all engine parameters indicating normal. Once airborne and indicating 60 mph, the Stearman started a gentle climb and a slight left turn to clear trees at the southeast end of the field with the wooden prop biting into the air. As the craft cleared the top of the trees, Jordan noticed that the airspeed was bleeding off. She rechecked mixture and power levers full forward. However, the airspeed continued to decrease, so she lowered the nose slightly to regain speed. As the speed continued to decrease, most likely from a slight tailwind or downdraft, Jordan lowered the nose to maintain above stall speed. With rising terrain ahead and no increase in performance, she elected to make a precautionary landing in an unpopulated open field located just northeast by ½ to ¾ mi of the departure strip. (ASRS 426347)

Those tricky winds

This is truly flying like the old barnstorming days—an uncontrolled dirt field with a windsock and a Stearman. The situation as described requires a lot of the pilot. There is no tower to help and no weather station on the field. Since the field was surrounded by trees, the pilot could not be totally sure of the wind conditions prior to takeoff. The windsock did indicate calm, but 60 ft above is another story. These factors called for conservative planning.

The pilot did not know the true conditions of the wind. However, the pilot did know she was using a wooden prop, which is nostalgic but also less efficient than a metal prop. With these conditions in place, the pilot still elected to take off with a full load of fuel. Extra fuel means extra weight, which in turn decreases performance. When the aircraft encountered the tailwind and/or downdraft, the pilot realized she didn't have the required performance. In situations like this, be conservative. A helium balloon works nicely to discover wind direction above the trees. Plan your fuel and your takeoff weight to give you a safety margin of error.

Carrying the Mail—Quickly

It was one of those overnight delivery companies out of Memphis. Julie, the captain; Jeff, the first officer; and Ray, the second officer, were winging westward in the 727. Originally, they were scheduled Memphis–Colorado Springs–Grand Junction, Colorado (MEM–COS–GJT). Colorado Springs weather was below minimums and forecast to remain that way. The flight was dispatched from Memphis directly to Grand Junction. Jeff did not notice the "non tower ops n/a" (not authorized to do non-tower operations, must wait for tower to open) message on the flight plan release. Because of the Colorado Springs overflight, the plane was scheduled to arrive at GJT 20 min prior to ATC Tower opening. The crew became aware of this fact when the company SELCAL'd (selectively called over the radio) it about 15 min after departure. Company desk personnel said to "fly slow or take the scenic route" until the flight could get authorization from the duty officer for a non-tower landing. The crew was told to expect a SELCAL around Colorado Springs giving it authorization to land at Grand Junction before the tower opened. The SELCAL came around 20 mi south of Colorado Springs.

The captain, Julie, was surprised when the flight controller said the weather at Colorado Springs was now above minimums and she could go there. Surprised, because she had just listened to ATIS about 5 min earlier and it said weather was still ¼ mi visibility (below minimums). Julie, Jeff, and Ray talked directly to Colorado Springs Tower, and weather was indeed above minimums, so they asked Center for a reroute (they were only about 15 mi away at this time, FL280). The 727 was cleared to Black Forest VOR with a descent to 9000 ft. They were all extremely busy. Jeff was flying, and Ray, the second officer, was copying divert information and arrival weather and calculating landing performance. Julie was getting out Colorado Springs plates and briefing the approach (ILS Runway 17R). They decided to conduct a monitored approach as the weather was just above Category I minimums.

Jeff was descending fast with speed brakes out. As the crew approached 9000 ft MSL, Approach gave it an intercept vector for the ILS and said to maintain 9000 ft until established. Jeff thought he heard 8000 ft and queried Ray. Julie was talking to Approach and did not hear him. Julie wears the Sennheiser headset and it is very quiet, but it sometimes makes it hard to hear cockpit conversation and aural alerts. Jeff continued the descent below 9000 ft. Julie did not hear the aural alert as they went below 8700 ft. She was tuning the ILS as the aircraft was in a left-hand turn to the intercept heading of 140. Abruptly at 8500 ft, Approach came on and said, "Don't descend below 9000 ft until on the localizer." The first officer immediately applied go-around power and raised the nose rapidly. The speed brakes were still extended and the stick shaker activated (speed was 10 knots above

0-degree bank minimum maneuvering speed). The captain asked the first officer to roll wings level, which he did. The speed had decayed below 0 degree minimum maneuvering speed, so Julie said she was going to flaps 2 degrees. The in-flight warning horn sounded because of speed brake extend/flaps extend. Jeff immediately retracted the speed brakes, and they promptly climbed back to 9000 ft. At this point Julie asked Approach to vector them back around to initiate another approach. This they did and completed the approach and landing without further incident. (ASRS 419560)

Making time

This case study is also full of valuable lessons, many of which have been previously covered. First, the crew exhibited poor planning prior to takeoff by missing the message concerning "non tower ops n/a." This is a planning error, pure and simple. Further in the report the crew relays that fatigue was a factor. It often is during night flights. Those of you on the "freight hauling, mail carrying" circuit know what it is like to try and sleep by day and fly by night. Fatigue will be discussed in the next chapter, but the best remedy I know is to try and adjust your schedule so that you switch your days (sleep) and nights (stay awake) as much as possible. This may be difficult if you have a family, but it does work.

This crew went from killing time (taking the scenic route) to making time (getting down now). It falsely perceived that since it was only 15 mi from Colorado Springs, it had to land immediately. Did the crew have to land so quickly? The answer is no; if anything, the weather would have gotten better as the morning sun continued to burn off the fog at the airport. The only reason the crew thought it had to get down fast was that it perceived it had to get down fast. It is all in how it framed the problem.

Framing

The concept of framing helps decide how we will solve a particular problem. Research tells us that situations can be framed in terms of positives or negatives. When it comes to negatives, we seem to be risk taking. This is known as the "gambler's fallacy." For instance, a man goes to the racetrack and loses money in the first race. He can go home with a sure (but small) loss or remain at the track and bet on the last four races and hope to erase his debt. If he stays, he most likely will return home with even a larger loss.

When it comes to positives we are risk-aversive. If a working man or woman had $150,000 in his or her retirement account, he or she is much more likely to keep the account invested in a "safe" but lower-interest fund than in a speculative investment that offers big returns and big risks. The person already has a positive (a large retirement account), and he or she is averse to risking this positive.

This financial lesson applies here. The crew can frame its situation as a positive or a negative. Here is the negative frame: If the crew chooses to hold up at altitude and get everything together, it will burn fuel and time (a negative). But if it proceeds down, it may be able to land quickly (the riskier choice). With a negative frame, the crew is likelier to make the risky choice and proceed with a rushed approach.

However, the crew can frame this situation as a positive. If the crew stays at altitude, it is doing the safe thing (a positive). However, if it proceeds down to the airfield quickly, it may have an accident (the riskier choice). With this positive frame, the crew is much more likely to choose to hold. Remember, how you choose to frame the problem will often dictate how you choose to solve the problem.

What the crew should have done was to enter a holding pattern. To the west of Colorado Springs the Rampart Range mountains rise rapidly. The crew should have chosen to loiter east of Colorado Springs, which is known as "west Kansas" because it is so flat. Always consider the terrain when deciding where the best spot is to hold.

Once it was established in a holding pattern, the crew would be in an excellent position to get up to speed and "get their ducks lined up." It had clearly not planned to land at COS prior to takeoff because of the poor weather, and things hadn't changed en route until the last minute. It takes a few minutes to regear mentally, and screaming down on an en route descent is not the place to do so. Use the holding pattern to plan.

The equipment

We already devoted a chapter to equipment and one to automation, but these factors came into play here. The captain had a problem with her equipment; it was too nice. Her headset was so quiet that she was unable to hear cross-cockpit communication and the aural warning horn. It is not totally clear, but the aural warning horn was likely triggered by the first officer busting through the altitude set on the altimeter. Obviously, aural warning horns are only helpful if you hear them.

Two other pieces of automation helped the aircrew in this situation. As the first officer got slow in his turn and climb because of the extension of the speed brakes, he got the stick shaker. The stick shaker warns of an impending stall. The captain was wise in telling the first officer to roll out of the bank, which would immediately increase the lift produced by the wings and decrease the stall speed. The captain elected to extend the flaps to further increase lift, apparently not noting that the speed brakes were extended. That's when the horn sounded for simultaneous flaps and speed brake extension. This is a nice use of automation by the designers. A good lesson for aircrews is that if speed is bleeding off, there can only be a few explanations: power is not up, winds have changed, or some kind of drag must be hanging from the aircraft, such as flaps, speed brakes, or landing gear. In this situation everything was happening quickly, so the captain treated the symptom (slow speed) instead of looking for the root problem (extended speed brakes). Of course, the crew would have not have been in this position had it not chosen to rush down.

Poor CRM was at work as crew members failed to confirm altitudes with one another when there was a question as to the clearance. However, the crew finally exhibited good in-flight error mitigation (breaking the error chain) by taking it around for another approach. This gave the crew time to finally get things in order. I wonder if its decision to perform a rapid descent would have been different if it would have had a load of passengers instead of a load of mail. What do you think? Does our cargo drive our risk taking? Passengers certainly wouldn't enjoy a rapid descent of 19,000 ft in 15 mi. Usually we choose to triple our altitude to determine the number of miles out to start the en route descent. So, a 19,000-ft descent would call for a start of 57 mi. There is no answer to this question, but it is one worth pondering. Would it make a difference to you?

Make Room for the Emergency

"We were heading into Kalamazoo/Battle Creek (Michigan; AZO), the cereal capital of the world. We were in IMC, rain, and turbulence. Center had us maintain FL200 until 15 mi out from AZO because of traffic. It then cleared us to 2500 ft, along with about four greater-than-40-degree heading changes for descent. The ride was turbulent. When handed off to Approach, we were told to maintain 2500 ft until established on the radial of the VOR approach and to keep our speed up, cleared for approach. An airplane that had just departed had declared an emergency and was coming back to land because of the problem. About 6 mi prior to final approach fix, we were given a vector that put us past the inbound radial. Then we were turned back to intercept. At this point we were about 4 mi from the FAF at 2500 ft established on the radial.

"My first officer is also a captain on another type of aircraft that we operate and is very experienced and competent. During the approach, I was keeping the aircraft on heading and altitude *and relying on the first officer for an approach briefing.* After I was established inbound on the proper radial (VOR approach), the first officer called final approach fix and the minimum decent altitude, and we started down. As it turned out, he called the FAF early, and we were descending below the initial approach altitude before we should have. About 1 mi from the actual FAF, Approach told us to climb; we had descended below 2500 ft to approximately 1800 ft. After establishing ourselves back onto the proper altitude, I reviewed the approach plate and discovered the error. The rest of the approach was uneventful.

"We debriefed later and discussed proper procedures. Basic procedures need to be reviewed on a regular basis. Otherwise, old lessons learned can be forgotten—things such as always accomplish a proper approach briefing, even if you are flying with someone that you trust and even if it means delaying the approach; and follow your standard operating procedures—both pilots should properly review critical items. I think the problem could have been avoided if we had not been left up so high so close to the approach given the weather conditions, and the rush because of the aircraft in trail with a problem." (ASRS 419645)

Nice debrief

This pilot does an excellent job of critiquing his own flight. Again, we see the well-established connection between rushing and poor planning. The crew may have fallen prey to the framing effect, figuring that the emergency aircraft may shut down the runway. It certainly would not be desirable to hold over a closed runway in the conditions described by the crew (a sure negative). However, if the crew framed it as a positive, "going into holding will give us a safe flight," it may have chosen a different course.

As flyers we like to establish and keep a good working relationship with ATC. It tries to help us out, and we respond in turn. However, this is a case of the crew complying, to its detriment, with a steep descent, many vectors, and then a rushed approach, all in order to accommodate ATC. You do have the power to say no. If you are being rushed, simply reply "Unable to safely comply." That phrase may not earn you points with the controller, but it will earn you more time as he or she gives you vectors back around to sequence you for another approach. In this case, the pilot did not even have time to personally review, let alone brief, the approach. I wonder what the pilot team had been doing at cruise altitude, when it should have been briefing the approach?

Discussion Height—I'll Take the Bus

"It was January 3, 1989, and I'd been a fully qualified Shorts Brothers C-23A aircraft commander for about one month. The C-23A, a very ugly camouflaged twin-engine turboprop, has a crew of three: the aircraft commander, the copilot, and the flight mechanic (a multitalented cross between a crew chief and a loadmaster). Our job that day was to haul 'trash' all over the southern United Kingdom and back to home base in Zweibrucken, Germany. Specifically, Zweibrucken to Mildenhall to Kemble to Mildenhall to Bentwaters to Zweibrucken. We logged a grand total of 6.5 flight hours that day. The weather in the United Kingdom that day was uneventful as I remember it. We were just MAC dogs flying low and slow over the United Kingdom enjoying the sights, and I was thinking about my *childless* trip to Paris the next day with my wife. We had sitters for the kids, and everything was set.

"In Bentwaters Base Operations, we reviewed the weather, flight plan, and slot time information for the leg back across the channel into Germany. At that time of year it was quite common for a cap of fog to form as the sun goes down at Zweibrucken AB because it sits up on a hill. A beautiful German valley lies below. The temperature drops rapidly to meet the dew point, and poof! A thin but dense layer of fog. We had a joke that in the eleventh month, November, the fog burned off at 1100. In the twelfth month, December, the fog burned off at 1200. No big deal, though—the current and forecast weather was to be above our PAR (precision approach radar) landing minimums. We'll just find a good alternate, put on the required fuel, and check it a few times en route to be certain the weather shop was right.

"As we checked the Zweibrucken weather en route, it seemed to be going down as forecast, but we were prepared so we pressed on. About 30 min out we radioed the squadron to get a parking assignment and to notify the cargo folks that we'd need a forklift. As an aside I asked the friend of mine on the other end of the radio what the weather teletype repeater at his desk said the current RVR was. It was still above minimums but going down quickly. The race was on! If we diverted, it meant we didn't get home that night, which would put a hitch in the trip my wife and I had counted on for so long. We would have to land at the alternate and take a several-hour bus trip back to Zweibrucken. Do you smell a classic case of get-home-itis brewing? I didn't.

"We briefed and set up for the PAR at Zweibrucken and asked that all the runway lights be turned up to their fullest intensity. The weather was clear to the stars, but we could see a 300 ft layer of clouds over the airbase. The last check of the runway visual range (RVR) was on the ATIS and it was right at minimums. Actually, it was starting to drop below mins, so my buddy thought he would help me out with a generous call on the RVR over the radio. I decided that our best shot at getting in was to not ask for the weather again and see how it looked at minimums on the approach. We started down final with the comforting words from the controller of, 'On course, on glideslope.' Still in the clear but the fog was nearing. At 100 ft above minimums we went from clear above to a very disorienting pea soup instantly. At minimums we were supposed to see the high-intensity runway lights, approach light system, or runway end identifier lights to continue to the runway. I reached decision height (DH). (The C-23 flies so slow the pilots actually called it 'discussion height' —so slow you have time to talk it over.) At DH, I don't see a thing, but thought, 'It must be there; the weather is right at mins.' I decide to count 1 potato, 2 potato. I thought I saw a light...maybe. We continued the approach. I saw a few more lights but didn't have much depth perception at all. I knew (hoped) the runway was right under me, so I chopped the power, cut the decent rate slightly, and slammed onto the runway. Heart still thumping and no one screaming; we must have made it.

"The next challenge was to taxi to the ramp. The visibility was so bad that I couldn't find a taxiway to exit the runway, and they were lit. I had never seen the fog that thick before. The usual 10 min taxi took me 30 min to complete.

"God let me off easy that night with the stipulation that I learn a very important lesson and pass it on. I should not have let my personal desires for a Parisian vacation interfere with my professional responsibility as an aviator and especially as the aircraft commander. I intentionally played ostrich and downplayed events that screamed DIVERT. I made a huge error, learned from it, and will not make the same one again. I learned my lesson. Next time I'll take the bus." (personal communication)

Get-home-itis

Get-home-itis is a topic we have already discussed. It is such a common problem that it bears repeating. This is a classic example of pressing to get home. All the factors are there. It has been a really long day. For this analysis I'll call our pilot Gary. Gary was very motivated to get home by factors far removed from aviation. He has a wife waiting to go away with him for a three-day trip to Paris without the kids. You don't know how long they have been looking forward to this. Maybe you do; if you have small children, you know how precious they are. You also know that you rarely get a break to be alone with your spouse during those baby and toddler years. Something like this came along for Gary and his wife maybe once a year or two. He didn't want to miss it. But if he had to divert, that bus ride back would cause him to miss his train to Paris. The trip would be dampened, if not ruined.

Of course, the alternative could be to miss the trip (and all other trips) because you are dead. Or you could miss this trip because you are explaining to the wing commander why you dinged up an aircraft. Or you may be visiting your chiropractor to have him reset your back that you threw out with the hard landing. Again, the remedy is very simple, but not easy. That is, it's simple to prescribe but hard to follow. Weighing the benefits of landing versus the potential costs of the attempt should encourage us to go around. Gary's decision of what to do was affected by the time pressure he felt to get down.

With friends like this, who needs enemies?

Gary's friend (Tim) at the squadron was trying to help him out. Some friend, huh? Almost got Gary killed. Tim is a classic example of unsafe supervision with good intentions. Tim's primary job is to advise aircrews as to the conditions at the base and be there in case they need assistance for emergency or other irregular situations. Tim's secondary role is to help out his buddy. How close was the RVR to minimums? We don't know for sure, but we do know that the RVR was less than minimums. Tim gave it a generous estimate. For those of you in supervisory positions, remember, your first duty to your friend up there is to give him or her honest feedback and then let your friend make the call. Resist the temptation to color things in a way that makes them seem better than they actually are. In the end, you aren't doing anyone any favors with that approach. The best thing you can do for any pilot is to give him or her an accurate picture.

The Hurry-Up Game

Get-home-itis, get-there-itis, or pressing has a very close cousin called the "hurry-up syndrome." NASA published an excellent report on the subject, authored by Jeanne McElhatton and Charles Drew, in 1993. The study was entitled "Time Pressure as a Causal Factor in Aviation

Safety Incidents: The Hurry-Up Syndrome." Though get-home-itis is often associated with general aviation, it can also affect commercial carriers, and those carriers were the target of the NASA study.

The hurry-up syndrome was defined as "any situation where a pilot's human performance is degraded by a perceived or actual need to 'hurry' or 'rush' tasks or duties for any reason" (McElhatton and Drew 1993). The study found that 63 percent of these types of incidents occurred during the preflight phase, and 27 percent occurred during the taxi-out phase. So 90 percent of these incidents occurred prior to takeoff. However, often the consequences of the incident were later felt in the flying portion of the sortie to include takeoff and climb-out. These consequences were sometimes killers. The consequences did not always result in injury or death. Deviation from company policy or procedure accounted for 20.8 percent of these incidents, while deviation from ATC clearance or FARs composed almost half (48.0 percent) of these incidents.

How does a pilot avoid the hurry-up syndrome? McElhatton and Drew make these excellent suggestions:

1. Maintain awareness of the potential for the hurry-up syndrome in preflight and taxi-out phases of flight and be particularly cautious if distractions are encountered in these phases.

2. When pressured to hurry up, particularly during preflight, take time to evaluate tasks and their priority.

3. If a procedure or checklist is interrupted, return to the beginning of the task and start again.

4. Remember that strict adherence to checklist discipline is a key element of preflight and pretakoff (taxi-out) phases.

5. Defer paperwork and other nonessential tasks to low-workload operational phases.

6. Keep in mind that Positive CRM techniques will eliminate many errors.

Many of these suggestions are common sense, but they are often not followed. It requires a great deal of self or flight discipline to slow down and do what is right.

Getting It Straight

Scheduling and planning go hand in hand. Proper planning is essential in order to cope with the hazards and surprises of flight. Improper planning can cause two different time-related problems. The first problem is delaying a decision beyond usefulness. A good example is Flight 173 into Portland where the captain insisted on several gear checks and on giving the flight attendants time to prepare the cabin for landing. The decision of when to head in for the approach became moot when the crew ran out of fuel and was unable to complete an approach or landing. Had the captain done some time planning at the beginning of the malfunction analysis, he would have known how much time he had to work with.

The second problem is that of jumping at conclusions prematurely. The freight haulers out of Memphis are an excellent example. They had intended to bypass Colorado Springs because of bad weather. The situation changed a mere 15 mi from the Springs. Instead of sitting back and assessing the situation, the crew elected to begin an immediate descent with arms and charts flailing.

Both of the problems are caused by temporal distortion, when time is either compressed or elongated. This in turn can be influenced by how we frame the problem. Our time judgments can become warped. When our time judgments become warped, we may make decisions prematurely or may delay them until they are no longer viable.

The take-home lesson from this chapter should be that planning on the ground will save you time and danger in the air. What you invest in planning pennies will come back to you in time-saving dollars. You can bank on it. If there is time to fly, there is time to plan.

References and For Further Study

Berg, S. L. and T. B. Sheridan. 1984. Measuring workload differences between short-term memory and long-term memory scenarios in a simulated flight environment. Proceedings of the Twentieth Annual Conference on Manual Control, NASA Ames Research Center, Moffett Field, Calif. pp. 397–416.

Kahneman, D. and A. Tversky. 1984. Choices, values and frames. *American Psychologist,* 39:341–350.

McElhatton, J. and C. Drew. 1993. Time pressure as a causal factor in aviation safety incidents: The hurry-up syndrome. Proceedings of the Seventh International Symposium on Aviation Psychology, Ohio State University, Columbus.

Miller, G. A. 1956. The magical number seven, plus or minus two: Some limits on our capacity for processing information. *Psychological Review,* 63:81–97.

11

Fatigue and Other Factors

Throughout this book we have highlighted the leading causes of CFTT and CFIT. There remains a few loose ends to tie up—factors that arise periodically that contribute to this problem. They do not occur frequently enough to warrant an entire chapter, but they do bear mentioning. This chapter will be a little bit like homemade vegetable soup, with a little bit of everything thrown in. However, many of these additional factors can be brought on by fatigue.

An All-Nighter

The following crew had the joy of an all-night flight. Listen as the captain describes what occurred.

"We departed Narita, Japan, en route to Anchorage. It's an all-night flight—5 h, 50 min. We were asked to keep our speed up on descent to accommodate a flight behind us. The first officer was flying our wide-body jet. Below 10,000 ft we were given vectors away from the direct track to the airport to increase spacing with the aircraft in front of us. We were cleared to 2000 ft and had the aircraft ahead of us in sight, but there was low fog in the area below us. We were then vectored toward the localizer. At 4000 ft, I said '2 to go' (2000 ft), and at 3000 ft, I said '1 to go' (1000 ft). We were then cleared for the approach, and I said to the first officer, 'Now we can go down to 1600 ft,' the initial approach altitude to Runway 6R at Anchorage. The second officer and I were then focused on the traffic ahead, and when I looked at the altimeter, we were at 1400 ft and descending. I said, 'Watch your altitude; we're too low.' We were about 12 mi west of the airport. The first officer corrected back to 1600 ft, but in the process we descended to about 1250 ft before the descent was arrested. Approach Control called and said to check our altitude as the first officer was already correcting back. A normal approach and landing was subsequently accomplished. The RVR (runway visual range) was 2400 ft and the wind was light and variable. A human error because of inattention and being up all night contributed to this mistake." (ASRS 420456)

Squeeze the day

Jet lag makes for tired bodies. This crew was going in the toughest direction: west to east. Our circadian rhythm and internal clock wants to stretch if given the opportunity. Research on folks who volunteer to live in caves for months at a time indicates that the human body prefers about a 25- or 26-h day. Therefore, traveling east to west and making a longer day is easier on the body than traveling west to east. West-to-east travel shortens the day.

Set up for failure

This crew was set up for disaster by ATC. The crew members had a lot of factors to contend with besides tired bodies. For sequencing reasons the crew was given a rapid descent to accommodate the aircraft behind it, which made for a time rush. Then the crew was given some irregular vectors to allow for proper spacing with the aircraft in front of it. The rush and then the slowdown disrupted habit patterns. I once heard an experienced flyer say that habit patterns can either save you or kill you in a crisis. This crew was not yet in a crisis situation, but it was heading for one.

Fatigue, the rapid descent, and the funky vectors led the crew to lose its attention on what was important. It became fixated and distracted by the aircraft in front of it and the low-fog conditions. Fortunately, the captain practiced some good CRM by cross-checking his first officer on his altitude. Some of the most consistent results of fatigue are inattention, distraction, and channelized attention. There are other consequences as well, as we will see in the following story.

Some Noise Complaints

It had been a long day. The DC-10 crew was already on a 12-h day after flying 5.5 h from Mumbai, India. Then it had a ground delay for cargo loading, a runway change, and an ATC delay for its route of flight. Furthermore, it had several time zone changes the previous day. The crew was fatigued and ready to get the last leg of the day out of the way. The DC-10 was at the hold line. Gary was the first officer and was flying the Perch 1 Standard Instrument Departure (SID) from Hong Kong International to Mumbai. The SID called for a runway heading of 073 degrees to point PORPA, then a right turn to point BREAM. The turn was to be initiated at 5 DME off of Hong Kong localizer, frequency 109.3.

Shortly after takeoff Gary asked the captain, Greg, to activate the FMS for the SID by selecting "heading select." Greg did, and Gary followed the course needle.

Shortly thereafter, Gary read 5 DME on his VOR and expected a right-turn command from the FMS to take him to point BREAM. It did not give him the command, so he manually selected BREAM direct. As he started his right turn the captain was busy changing the radio frequency to departure frequency. After contact they were given an altitude of 7000 ft, which the captain put into the FMS and the first officer confirmed. At this time the controller asked if they were still on the SID. Gary replied, "Yes, we are right on it." The controller then said, "You are right of course and approaching high terrain." At this point the captain took control of the aircraft and immediately started a climbing left turn. At the same time he commanded the engineer to push the throttles up to takeoff power. They then got a GPWS warning. Shortly thereafter the controller said they were clear of the terrain and cleared to point PERCH (the next point after BREAM).

After it was safely airborne the crew researched the problem. Gary discovered that he had set the wrong navaid in his VOR, causing it to read 5 DME much sooner than required for the SID. The problem was discussed with the controller. He said he had no problem with it, but he did receive some noise complaints. (ASRS 427106)

Out of your zone

Fatigue can cause us to miss the little things. This crew was understandably fatigued after 12 h this day and several time zone changes the day before. Jet lag can bring on fatigue and wear us down. There are no foolproof ways to combat jet lag, but there are few things we can do to help protect ourselves from the adverse effects. The first is to maintain good physical conditioning. This will help us remain more alert with or without a time zone change. The second is to watch our diet, which is always difficult on the road. Eating while on a trip too often becomes a quick breakfast, the same old thing on the plane, and a late dinner with meat and a starch. We can only run on what we put in our engine. Another tip is not to sleep heavily just after landing on a trip. Try to stay awake or take a cat nap of less than 20 min. Force yourself to stay up until about 10 h before your next flight and then hit the sack for at least 8 h of sleep. Finally, go light on the caffeine. Caffeine can make time zone changes more difficult. I realize that these suggestions may be difficult to follow, but they do help if you can implement them.

Missing the little things

The first officer made a simple little mistake. We covered the importance of properly programming the FMS in Chap. 6. Remember, if you put garbage in, you get garbage out. In this case Gary had programmed the FMS correctly but dialed the wrong frequency into the navaid. Gary actually showed some initiative in starting the turn thinking that the FMS had failed to make the turn. He was not blindly following the automation. However, he was blindly trusting the navaid. It was unfortunate that the captain was on the radio at this time. He could have provided some backup and helped to determine why the FMS had failed to initiate the turn. The lack of a cross-check allowed the error to continue. The first officer was busy flying, so he didn't take the time to really analyze what was going on.

The captain exhibited some fine CRM in this instance by grabbing the yoke upon the warning by ATC and the subsequent GPWS alarm. He took quick and decisive action, which was justified, and turned a CFIT into a CFTT.

Reserve Duty

It was a normal day on reserve. Laura got up at 0830. She was notified about the trip at 1030. The trip had a deadhead to Chicago O'Hare (ORD) at 1200 midnight. After finishing some things she was doing, she tried unsuccessfully to take a nap and left her house at 2230 for the deadhead flight. She slept a little on the flight to ORD. The crew had a 3-h, 46-min wait in ORD, so she tried to sleep there for an hour, but only dozed. Then the crew flew to Anchorage (ANC), a 6-h flight in the 747-200. Laura had only had a little fitful rest in the last 27 h. The other crew members had similar stories in terms of rest. The Anchorage ATIS stated that the ILS to Runway 6R was the approach in use. Laura briefed that approach. Abeam the field on downwind, the crew was told that Runway 6R was closed for plowing and the approach would be flown to Runway 6L using the Localizer Runway 6L approach. She asked the pilot not flying to ask for 10 mi so the crew could brief the new approach. She was not specific enough, and instead of an extra 10 mi, the crew got a 10-mi final. She briefed the approach but felt rushed and did not mention a visual descent point. If there had been a published visual descent point (VDP) on the approach chart, it would have helped the crew members, since they are trained not to violate a published VDP unless needed to make a normal landing. She didn't brief it and there was no reminder on the approach plate, so she wasn't thinking VDP.

So what did she do? Exactly what she does in the simulator. At the MDA and seeing the approach lights, she started a slow descent for landing. The problem was it was too early, and the plane got within 100 ft AGL, 1 mi from the runway (very low). The first officer called "100 ft," and the captain added power to raise the nose and gain some altitude to get back on a proper descent profile before seeing the VASI and correcting. The MDA was only 339 ft AGL, so there was little room to descend below MDA and still be safe. In hindsight, all the crew had to do was use the available DME and compute its own VDP, which would have been about 3 DME. (ASRS 429116)

Bad transfer

Earlier we spoke of negative transfer of training. This is where we take something well ingrained during training that doesn't quite fit the real world. Yet because we have rehearsed it so many times, we fail to recognize that it is not the correct response in the actual situation. The crew members had been practicing nonprecision approaches in the simulator that demanded they descend as soon as they hit the MDA in order to make the approach. That led the crew to leave the MDA at Anchorage as soon as it had any indication of the runway environment. Unfortunately, those indications were sketchy so the crew ended up doing a 100-ft low-level flight 1 mi from the runway. One hundred feet in a Boeing 747 is a little "burp" from the ground. That is why the VDP is so essential. The crew member makes a good suggestion to include the VDP on published approaches. Those approaches with this feature serve as a good reference and a good reminder to calculate the VDP appropriate for the aircraft's optimal glideslope.

The captain's comments

The reporter had the following interesting comments to make about the incident. "Did fatigue play a factor? Sure. Do we have a policy where we cannot fly if we are too fatigued? Sure. The problem is making the right decision before beginning a 6 hour flight. Once airborne, it's too late. Please push the charting authority to add a VDP to the Localizer Runway 6L approach plate at ANC. I didn't compute and use a visual descent point (VDP) and almost paid a very dear price for that omission. I know we are told to use all available aids during the transition from approach to landing phase, but I still suggest re-emphasizing it. I know better and I usually do compute a VDP and even have several indicated on charts in my flight kit....Why didn't I this time? Part of that answer is...fatigue. I won't speak for my other crew members, but I know that my mind was slow and so was my cross-check."

The captain does a nice job of suggesting what went wrong. Those who have lived the airline life have suffered life on reserve. It is tough to get the proper crew rest on reserve. It always has been and it probably always will be. As the captain points out, there is a company policy to protect pilots, but the tough part is making the call before the flight. Once airborne, it's too late. Calculating a VDP is not required, but it is extremely helpful when you are flying a nonprecision approach. This is an accepted technique. This crew did not calculate the VDP for two reasons. One is that it was fatigued. The other is it was rushed by ATC. Poor communication between pilots and ATC turned a good suggestion (extend the approach for 10 mi to help us prepare) into a poor outcome (flying a 10-mi final unprepared). The captain should have reiterated her actual request to the first officer. The message doesn't always get across the first time. Clarify it until ATC gets it right. However, fatigue gets us to settle for less-than-optimal performance, which may jeopardize our safety. The captain wasn't willing to pursue the issue as she should and later she had regrets. In this case, fatigue caused the captain to settle for the 10-mi final instead of the 10-mi extension—that wasn't healthy.

From these cases we have seen that fatigue causes us to settle for less-than-optimal performance. It slows our cross-check by impairing our reaction time. It leads to irritability, fixation, and channelized attention. Fatigue can also make us susceptible to other factors hazardous to flight, as we will see next with a young A-4 pilot.

Twinkle, Twinkle, Little Star

Don took off in his A-4 from Yuma Marine Air Station. It was the 1960s and there was nothing, I mean nothing, outside of the Yuma city limits. It was "clear and a million"—probably 150 mi visibility at altitude. He was off for the range to dogfight and drop bombs (and play), a totally VFR mission. Coming back to Yuma, he could see the runway about 50 mi out. It was so clear he can see the individual runway lights. It was one of those evenings when you can't believe they are paying you to fly! Suddenly, the lights around the runway begin to twinkle and come in and out. It seemed funny to Don. "There must be a haze deck down there," he thought. It was still very clear, but the runway lights continued to come in and out and twinkle. He continued to fly toward the base—40, then 35, then 30 mi. Suddenly, he noticed something black. It was a ridge line. The in and out was caused by going slightly below and then above and then below the ridge line. The twinkle came from the glare of the runway lights off of a stretch of rocks on the top of the bare and dark ridge line. He looked down at the MEA; he was below it. It was so clear he thought he was actually closer to the base than he was and had des-cended too quickly. (personal communication)

Illusions of grandeur

Aircraft flight can be fraught with visual illusions. Many of these illusions are beyond the scope of this book. However, most illusions occur at night, when fatigue is also a factor. Furthermore, night vision can deteriorate with age. It can also be adversely affected by smoking tobacco. There is nothing you can do about the age thing. There is only one alternative to having another birthday. However, laying off of tobacco will increase your night vision.

Don was lulled into a dangerous situation. It was so beautiful and so clear that there seemed to be little to worry about. So clear in fact that attention to the instruments didn't seem important. How often can you see an airfield over 50 mi away with nobody else around? Unfortunately, there is often deceptive terrain around unpopulated areas. This terrain is often unlighted because of its remote location. Such was the case here as Don slipped below the minimum en route altitude while channelizing his attention on the runway in the distance and admiring the beauty of the night. The lesson to take away is no matter how beautiful it is, you can't leave your instruments out of your cross-check. It is especially important for the altimeter and ADI to be in the cross-check at night.

Visual illusions

Visual illusions can result in CFTT or CFIT. Following is a list of several illusions to be aware of:

Light confusion

A common problem associated with night flying is the confusion of ground lights and stars. Many incidents have been recorded where aircraft have been put into very unusual attitudes to keep dim ground lights above because they were misinterpreted as stars. This normally occurs during a time of reduced visibility caused by fog, haze, or low clouds. Pilot have confused highway lights with runway approach lights or a line of ground lights with the horizon.

False vertical and horizontal cues

Sloping cloud layers may falsely be interpreted as being parallel to the horizon or ground. Momentary confusion may result when a pilot looks outside after prolonged cockpit attention. Night-flying techniques should place less reliance upon less obvious outside references. Disorientation can also because by the Northern Lights.

Relative motion

During ground operations, an adjacent aircraft creeping forward can give the illusion of the observer moving backward. Pilots may instinctively apply the brakes and even look at the ramp if the illusion is severe. Such illusions are also possible during formation flying.

Autokinesis

Another illusion of night flying is *autokinesis*, the apparent movement of a static light source when stared at for relatively long periods of time. Autokinesis will not occur when the visual reference is expanded or when proper eye movement is observed during visual scanning. To preclude autokinesis, aircraft have several staggered formation lights rather than a single light source. Illuminated strips have been installed on many aircraft to replace formation lights and are quite effective.

Haze and fog

Dust, smoke, fog, and haze tend to blur outlines and to reduce light intensities and colors of distant objects. These atmospheric conditions result in an apparent increase in distance of the object. Excessive haze or fog will cause a runway to appear much farther away.

Space myopia or empty vision field

At high altitudes or during extended over-water flights, pilots may develop a physiological adjustment of the eye because of a lack of distant objects on which to fixate. A varying degree of relative nearsightedness develops. For example, a pilot with normal visual acuity of 20/20 is able to discern an aircraft having a fuselage diameter of 7 ft at a distance of 4.5 mi. When accommodated to relative nearsightedness, the pilot would not be able to detect the same aircraft at a distance greater than 3 mi.

Night landing problems

Reduced visual cues received by the pilot during night and weather landings complicate distance judgment. There is always the danger of confusing approach and runway lights. When a double row of approach lights joins with the boundary lights of the runway, pilots have reported confusion in determining where approach lights terminate and runway lights begin. The beginning of the runway must be clearly shown. During an instrument approach in a gradually thickening fog, the pilot falsely senses a climb and in compensating may descend too low. Under certain conditions, approach lights can make the aircraft seem higher within a bank than when level. Pilots have also reported the illusion of the bank if one row of runway lights is brighter. Another illusion may occur if a single row of lights is used along the left side of the approach path. A pilot may misinterpret the perspective and correct to the left to center lights, causing his or her touchdown point to be too far to the left of the runway centerline. Instrument approach systems, combined with a standardized improved approach lighting and glideslope system, should eliminate or drastically reduce illusionary information.

Runway width

Illusions may be created by a runway that is narrower or wider than the runway to which the pilot is accustomed. On a narrow runway, the pilot may perceive a greater-than-actual altitude and could touch down short or flare too low. The illusion may be reversed when approaching a wider runway. The illusion is not as severe when other reliable visual cues are available and is easily overcome with a visual approach slope indicator (VASI).

Runway and terrain slope

Most runways are level, but those that have some degree of slope provide illusory cues to the pilot. For the up-slope runway, the pilot may feel a greater height above the train, causing a normal flight path to seem steep. Executing a compensatory glide path would result in a low final approach and an undershoot of the runway. A similar situation occurs when the runway is level but there is downslope in the approach terrain. The downslope runway provides illusory cues that indicate lower-than-actual height. A compensatory glide path would result in overshooting the runway. Upslope terrain in the approach zone creates a similar illusory effect. If a pilot is in doubt concerning the runway and approach terrain slopes, an instrument approach should be flown. Multiple illusory cues are usually encountered instead of the simplified single cues. Combinations of these cues may decrease or increase the total illusory effect. A severe illusion might result from a narrow upslope runway and a downslope approach terrain. Conversely, an upslope runway could be concealed as an illusion by upslope terrain or a wide runway or both. Inclement weather and poor light probably increase the effect of these illusions, and an instrument approach would again be a wise decision.

Additional landing problems

Over-water approaches minimize visual cues and contribute to low approaches and undershooting. Variable-height objects on the final approach may provide false cues to elevation. The best means of eliminating or reducing aircraft accidents from these types of illusory cues is careful preflight research and briefing. A pilot must be aware that judgment can be affected by changes in unusual visual cues and should fly an instrument approach if available. If no instrument approach is available, a low approach should be carefully flown before landing.

The role of the organization

Throughout this chapter we have seen the role of organizational factors in CFTT. The airline pilots on reserve were forced to deadhead to Chicago O'Hare and then make a long flight to Anchorage to land in difficult weather conditions. Their approach was shot near minimums after very little sleep over the last 27 h. That was simply one of the downsides of sitting reserve. Though there was a company policy to ensure proper crew rest, it was difficult to put into practice, as operational constraints forced the "adequate rest" call to be made long before the actual test of how fatigue had affected the crew. This is an example of how organizational factors can influence CFTT.

In Don's scenario above, his unit emphasized the importance of VFR flight as the squadron practiced their dogfighting and bombing runs. This emphasis carried over to the return leg to Yuma as Don marveled at his ability to see the runway over 50 mi away. The instrument crosscheck was dropped as his altitude dropped. Organizational influence upon operational practice is strong.

Finally, recall the T-38 low-level accident in Chap. 3. The squadron emphasized staying ahead of the time line. Of course, it also emphasized safety. But the time line was much more salient. It stood out. Accidents are periodic and rare. The time line is briefed daily to the squadron and wing leadership. That is what gets people's attention. Therefore, the young instructor pilot likely felt some organizational pressure to get that low level in.

The aforementioned studies set the stage for the next case study, which highlights an excellent example of the power of the organization in CFIT. The case is examined in depth by Daniel Maurino and colleagues in their book *Beyond Aviation Human Factors*. The case illustrates how latent organizational factors can manifest themselves in an operational aircraft accident. What is interesting about this case is that two separate accident investigations were conducted. Each investigation reached a different conclusion. Neither report can totally refute the other.

Mt. Erebus

The flight was a planned sight-seeing tour of Antarctica by a DC-10 flying from Auckland, New Zealand. Ian was in the left seat and Keith in the right. The company had been running the sight-seeing trips for a number of years, though Ian and Keith had never flown one of them.

The United States military has a base in Antarctica called McMurdo Station. The United States Navy Air Traffic Control Center (Mac Center) operates from this station. The flight was planned as a flight down to Antarctica, a descent to look around, and a return to New Zealand. It was a long trip, over 2300 mi one way, and no landing was planned at Antarctica.

The area for sight-seeing was reportedly overcast, with a cloud base of 2000 ft. Some light snow was falling, with visibility in excess of 40 mi. Not too far from the sight-seeing area, some clear areas were reported with a visibility of 75 to 100 mi. At roughly 40 mi from McMurdo, the crew spotted some breaks in the clouds that would allow it to descend down below the cloud deck. It requested and received permission from Mac Center to descend. It disconnected from the automatic navigation tracking system (NAV track) of the Area Inertial Navigation System (AINS) and manually descended in an orbiting fashion down to 2000 ft. Once it arrived at 2000 ft the crew decided to continue down to 1500 ft in order to give the paying spectators a better view. Once established at 1500, the crew reselected the NAV track feature so that the aircraft would automatically follow the computerized flight plan. It was now inbound from over the ocean to the main viewing area on Antarctica. Approximately 3 min later, the GPWS sounded. Fifteen seconds later, in a pull-up maneuver, the aircraft impacted the side of Mt. Erebus, elevation 13,500 ft.

Two conclusions

As mentioned previously, there were two separate accident investigations, which reached two different conclusions. The original aircraft accident report stated, "The probable cause of this accident was the decision of the captain to continue the flight at low level toward an area of poor surface and horizon definition when the crew was not certain of their position and the subsequent inability to detect the rising terrain which intercepted the aircraft's flight path." A subsequent commission of inquiry stated, "In my opinion therefore, the single dominant and effective cause of the disaster was the mistake by those airline officials who programmed the aircraft to fly directly at Mt. Erebus and omitted to tell the aircrew." It continues, "That mistake is directly attributable, not so much to the persons who made it, but to the incompetent administrative airline procedures which made the mistake possible." So we have two different conclusions by two different bodies. These conclusions will be more understandable as you read the causes behind the crash.

Organizational factors

How could this crash have occurred? There are a number of factors. First, the crew was not alerted by any regulatory authority or the company as to a phenomenon peculiar to the polar regions known as *sector whiteout*. This is not the same phenomenon known to skiers. In sector whiteout, visual perception is altered under specific cloud and light conditions. Sector whiteout makes white obstacles even a few yards in front of the aircraft invisible. This even occurs in nominal visibilities of up to 40 mi. Sector whiteout nullifies the protection normally afforded by VFR flight. The crew was not warned in any matter about this condition—a condition it had never seen. It was not included anywhere in the route briefing materials. The crew had no idea of the danger in attempting to fly visually over textureless white terrain in overcast polar conditions. It likely didn't realize it was even in such conditions.

The pilots were briefed on the route of flight. It was to run from Cape Hallet down the center of McMurdo Sound to the Byrd reporting point. This route had been flown by seven sight-seeing flights over the past year. The route was 25 mi to the right of Mr. Erebus. Six hours prior to the flight (during the night) the company's navigation section changed the last computer flight plan waypoint from the Byrd reporting point to Williams Field. This placed the route of flight directly over Mt. Erebus. Neither pilot was briefed of the change. So when they coupled the NAV track, they were essentially signing their death warrant.

The briefing given to the pilots included two minimum altitudes. One was 16,000 ft, which allowed clearance over Mt. Erebus. The other was 6000 ft, for use in a sector defined around McMurdo. There was confusion among company pilots on what these minimum altitudes actually meant. Many of the pilots felt that flight below these altitudes was permissible if VFR conditions prevailed. Whether they were clear minimums was debatable and illustrates a lack of clear communication within the company.

The captain and the first officer had never flown to Antarctica before. According to company planning procedures, a pilot must have a familiarization ride over someone's shoulder before he or she is permitted to act as pilot in command of the sight-seeing flights. This policy was eventually discontinued. It was felt that the briefing provided to the pilots was adequate to provide all the information needed for a safe flight.

One of the ground-based navigation aids, McMurdo NDB, was briefed as unavailable to the crew. The briefing officer had tried unsuccessfully to ascertain the actual status of the NDB. The Navy had formally withdrawn the NDB from use; however, the Navy decided to let it transmit until it failed. It indeed was transmitting the day of the accident. Without the NDB, there was only one operating ground-based navigation aid—a military navigation aid transmitting distance-measuring equipment (DME) information only. Again, a lack of communication was responsible for incomplete knowledge on the part of the aircrew.

The U.S. Navy had informed New Zealand's civil aviation authority that ATC would be provided in an advisory capacity only. In other words, it would be in a limited role. However, the company crews had been briefed to deal with McMurdo ATC as they would any other ATC facility. Crews normally rely on ATC to help ensure ground clearance, though it is a backup, not a primary source of this knowledge. However, if ATC service is limited and especially if this is not known by the crews, an aircrew becomes particularly vulnerable.

Crew factors

There were a number of questionable decisions on the part of the crew. It proceeded below two minimum altitudes without having a firm fix on its position. It did not use available ground aids to help in its descent. It attempted to remain visual and clear of terrain in whiteout conditions. According to cockpit voice recordings, the crew had some reservations on its ability to remain clear of the terrain as it proceeded toward the crash area. It took no action when it was unable to contact Mac Center on the VHF radio. This inability to connect suggests that a large obstacle (like a mountain) may have been between the crew and McMurdo. The crew did not use its inertial navigation system to cross-check whether its present course was taking it to the Byrd reporting point across McMurdo Sound. The crew seemed to assume that since it had good lateral visibility, the field of white in front of it contained a flight path suitable for VFR flight.

The biggest lesson that we can take from the Mt. Erebus accident is that organizational as well as flight factors can play into CFIT. Daniel Maurino and colleagues point out that there are several layers of defense in accident prevention. They begin with the organization and the policies and culture established by that company and continue through the planning and briefing stage. Once a plane is in the air they include air traffic control. Finally, the aircrew is always the last line of defense. When that line fails, usually the crew gets the blame—that is, pilot error. What should be clear from this case is that a whole series of events was set into motion before the DC-10 ever impacted Mt. Erebus. The crew's failures were simply the last in a long line of failures. This crew was set up by the organization.

There are often latent factors, factors below the surface, that are manifested only by aircrew accidents. Let me use football as an example. We usually pay attention to statistics such as yards rushing and passing. The number of completions, number of tackles, and number of sacks are all listed in the game's box score. But the latent factors often determine performance. The latent factors that underlie a football player's performance are essentially motivation, speed, and strength. The players with the most yards, tackles, and so on are usually the fastest, strongest, and most motivated players. It is these factors that determine to a large degree the performance of a football player. In the same way, latent organizational factors often determine aircrew safety and the prevention of CFIT. The latent factors in this accident include poor communication and supervision by company personnel toward the mishap aircrew.

Error Chains

The Mt. Erebus incident is yet another example of the error chain. As mentioned in an earlier chapter, an error chain is a series of errors that taken in isolation are not of great significance. However, when these errors are linked, they can lead to fatalities. As with all error chains, if the chain had been broken anywhere along the line, the DC-10 crew would not have crashed into the base of Mt. Erebus. So the key to error chains is breaking them.

How do you break an error chain? The answer is what some call a "red flag." Have you ever been talking to someone, perhaps someone younger, and he or she is seeking your advice? The person tells you his or her assessment of the situation, and something the person says doesn't sound right. It raises a "red flag" in your mind, indicating that you see trouble ahead. Extend that now to the aircraft. Something needs to trigger in your mind that danger lies ahead. That something needs to catch your attention, and you need to stop, realize that something is wrong, step back, think, analyze, and then correct. A good example comes from the pilot back in Chap. 3. He had taken off over the San Francisco Bay and headed for the San Mateo Bridge, which he thought promised VMC. Soon, though, he popped into IMC, and he was not instrument-rated. He began a descending turn toward what he hoped was VMC as the Bay drew closer. Finally, at 400 ft, he stopped the descent and counted to 10. He realized something was wrong; he stopped and regained his composure and then started a gentle climb away from the water. Stopping and counting to 10 was the red flag that broke his error chain. The Righteous Brothers sang of an "unchained melody." The error chain is a chained harmony. It is a chain that must be broken with red flags.

Stressed Out

Fatigue and organizational factors can lead to stress in the flyer. Stress, in turn, can facilitate the propagation of error chains and dull the pilot's ability to raise the red flag to stop the chain. For those reasons, the final factor we need to highlight is stress. Stress is that inner tension you feel. It can be brought on for the flyer by two different sources. There is *lifestyle stress* that comes from conditions outside of the cockpit. Things such as a move, a new job, a promotion, a poor performance rating, marriage, divorce, problems with children, credit card debt, sickness or death in the family, an argument with a spouse, a new car, and planning a much needed vacation are all sources of lifestyle stress. Notice that many of these things are good and desirable. However, stress can be positive or negative. Any stressful event can knock a pilot off of his or her equilibrium and affect his or her flying. Lifestyle stress can be distracting and long term.

The second source of stress is known as *flying stress.* It is brought on by conditions from the time you arrive for the flight until you go home for the day. It is often acute and short-lived. Examples include a disagreement with another crew member on a procedure, a fire light on engine number 4, and a missed approach because of bad weather. Stress carried over from one's lifestyle can impact what occurs in the cockpit. This sets the stage for flying stress to be more acute and severe than it may otherwise be. Things that we may normally brush off as the little stuff grows in size and can become a big problem.

Tracy Dillinger has written an excellent article on the different kinds of aviators. One is known as the Distressed Aviator. The symptoms of the Distressed Aviator include things like a sick child or spouse. You may have a family member with special needs. Do you feel like you are tasked with more than you can do? Are you involved in some required extra duty such as an accident investigation board? Do you feel burned out and unappreciated? If you said yes to a lot of these questions, you may be a Distressed Aviator. Before you get too upset with your diagnosis, you should know that this is frankly quite a common problem. It doesn't mean that you are a failure. These issues can be resolved with time and some creative problem solving. However, they do put you at a higher risk to miss a checklist item, to forget something for the flight, or to get sick. It is important to get some rest. If the Distressed Aviator doesn't take preventive action, he or she may slip into the Failing Aviator category.

For the Failing Aviator, life's situational stresses have become so overwhelming that normal coping methods fail to work. The normal coping methods include good health, good hand/eye coordination, a high need for mastery, and a very matter-of-fact way of dealing with stressors. The Failing Aviator loses his or her ability to compartmentalize the things in his or her life. They all begin to bleed together. This type becomes increasingly irritable and withdrawn. He or she becomes increasingly argumentative and engages in poor communication and CRM. Failing Aviators, of course, are in denial about their problem and their symptoms. They begin to experience sleep problems and may begin to abuse alcohol or tobacco. They also begin to violate flight rules.

What should be done to help the Failing Aviator? The first step is that someone must have the courage to identify the person with the problem. This can be difficult. Once he or she has been identified, that person should be taken off of the flight schedule. Give the person some time off to help him or her regain his or her composure. You can also seek help through the flight surgeon or refer the person to the chaplain or counseling services. Make sure the person has a friend or buddy to watch out for him or her. The goal is to get that flyer back up on his or her own feet. Without proper treatment, this flyer is a walking time bomb. It is essential for his or her welfare and that of the squadron to get the person some help.

Keep in mind, however, that not all stress is bad. An absence of all stress is known as death. Stress can actually key us up and help us focus on the task at hand. It can sharpen our minds. Stress and workload have a lot in common. Too little workload makes us complacent and bored. Too much workload overloads us, and our performance breaks down. We function best with a intermediate workload level. The same is true for stress; we operate best under an intermediate level of stress.

To determine the appropriate amount of stress (or workload), think of a gently sloping hill that rises to a short but flat plateau. At the far end of the plateau, picture a cliff that drops off almost vertically. That is the relationship between stress and performance. As stress increases, our performance improves in an increasing manner up until the point that it peaks or plateaus. We can maintain this level of performance for a small region of stress. However, if the stress becomes greater, our performance drops off rapidly as we are overwhelmed.

References and For Further Study

Dillinger, T. June 2000. The aviator personality. *Flying Safety*, 8–11.

Government of New Zealand. 1981. Report of the Royal Commission to inquire into the crash on Mount Erebus, Antarctica, of a DC-10 aircraft operated by Air New Zealand Limited.

Kern, T. 1997. *Redefining Airmanship.* New York: McGraw-Hill. Chap. 6.

Maurino, D. E., J. Reason, N. Johnston, and B. B. Lee. 1995. *Beyond Aviation Human Factors*. Aldershot, England: Ashgate Publishing Limited.

Office of Air Accidents Investigation, Ministry of Transport, Wellington, New Zealand. 1980. Aircraft Accident report No. 79-139: Air New Zealand McDonnell-Douglas DC10-30 ZK-NZP Ross Island, Antarctica, November 28, 1979.

12

Finale

Throughout this book and series, we have tried to build a framework for controlling pilot error. To err is human. To control error is to be an effective pilot. The first element in this framework is preparation. Proper planning prevents poor performance. Planning and preparation are applicable to more than just thorough flight planning. I am also referring to developing the proper attitudes, knowledge, and skills essential to flight safety. Proper preparation and planning can allow the pilot to avoid many of these outlined errors altogether.

The second board in this framework of error control is in-flight mitigation. How can we manage those errors that do occur in order to prevent an error chain from developing? Do you have the ability to raise the red flag and take action to compensate for errors that have been made in order to not only keep the situation from getting worse but to get the situation under control?

The final piece of the error control framework is post-flight analysis. Develop the habit of thoroughly analyzing errors that have occurred in flight in order to avoid those errors in the future. Learn from the mistakes that you will make. One helpful mnemonic is AOR, which stands for Action–Observation–Reflection. After the flight, analyze the erroneous portion. Determine the action and observe what the consequence was. Is the action something that you want to become habitual? After clearly outlining the action and observing the consequence, reflect upon the situation. How did it come about, how may it have been prevented, and what would you do the next time that situation arises? A thorough digestion of the flight errors can go a long way toward controlling future pilot error.

Swiss Cheese

Scott Shappell and Douglas Wiegmann recently completed a work that sheds valuable light upon human factors errors in aviation. Their work is based upon the pioneering work of "error guru" James Reason. Pilot error is implicated in 70 to 80 percent of all military aviation accidents. But as Shappell and Wiegmann point out, there are many levels of error. You can think of it as a system of four levels that produces errors. You can think of these four levels as four individual slices of Swiss cheese. When the holes in each slice line up, an accident occurs. (See also the Series Introduction.)

I will describe each layer of this Swiss cheese, working from last to first in my description. The last piece of cheese is the *unsafe acts* that the crew commits. Reason calls the crew "the last line of defense." When this line fails, an "active failure" occurs. Active failures are synonymous with a mishap. Examples of unsafe acts include skill-based errors such as a breakdown in the visual scan. Decision errors, such as a misdiagnosed emergency, are also unsafe acts. There can be perceptual errors caused by visual illusions. Finally, the crew can violate procedures or policies, such as failing to properly brief a mission. All of these types of errors and violations make up what the Swiss cheese model refers to as unsafe acts. The ASRS and NTSB reports we have reviewed have illustrated all of these types of unsafe acts.

The next piece of cheese is known as *preconditions for unsafe acts*. This level contains what are known as latent or underlying failures that have yet to manifest themselves. These include substandard conditions of operators—for instance, adverse mental states like channelized attention or get-home-itis. The pilot can also have an adverse physiological state such as physical fatigue. There can be physical or mental limitations such as limitations to vision and insufficient reaction time. Other factors include substandard practice of operators like poor CRM. Poor CRM can be manifested by a failure to back one another up and a failure of leadership. Finally, the personal readiness of the flyer can be jeopardized by self medication or violation of crew rest. Again, the studies we have covered have illustrated these concepts.

The next level of cheese is *unsafe supervision*. Inadequate supervision can be a failure to provide guidance. Operations can be planned inappropriately, such as a failure to provide correct data. Supervision can also fail to correct a known problem—for instance, unsafe tendencies like a pilot identified as an ogre. Finally, a supervisor can perpetrate violations such as authorizing an unqualified crew for flight. The Mt. Erebus accident outlined in Chap. 11 was a case study in failed supervision.

Our case studies have also illustrated the highest level of the system, *organizational influences*. Resource acquisition and management includes human resources, monetary and budget resources, and equipment and facility resources. The organization must be properly staffed and funded, and it must own/maintain suitable equipment. Furthermore, the organization establishes a climate or culture. The culture is propagated through the organizational structure, such as the chain of command and its ability to communicate. The culture can allow things like peer pressure to influence pilots to violate the regulations. Policies of hiring, firing, and promotion affect the organization's ability to operate safely. Norms, values, and beliefs are a part of any organization. They can run the spectrum from healthy to unhealthy. Finally, every organization has a process—from operations to procedures to oversight. The New Zealand sight-seeing company that operated the DC-10 flights to Antarctica had a climate marked by poor communication and a value of profit over following safety procedures.

I emphasize that each of these levels of the system plays a role in mishap prevention. The latent factors often go unnoticed in ASRS reports. We tend to focus on the unsafe acts portion of the system because it is most salient. However, many times there were failures throughout the entire system that led to the final aircraft mishap. We must be vigilant to examine all the levels of the system in order to control pilot error.

The Fundamental Attribution Error

One of the most powerful principles in psychology is known as the *fundamental attribution error*. This concept deals with the way that we assign (or attribute) blame (responsibility). It works like this: If you see someone do something that you find not up to standards, you are likely to attribute that action to a character flaw in that person. For instance, if you see a car swerve around a corner and run a red light, you are likely to say, "That person is reckless." If someone fails to meet a deadline at work or turn in his or her homework at school, you are likely to say, "That person is lazy." However, if the person who does this same deed is you, your explanation is likely to be very different. You are more likely to attribute the cause to the situation around you, rather than to your own character. For example, you swerve around the corner to miss the hole. You ran the red light because you had to pick up your child before dance lessons were over. You missed the deadline at work because you had other fires to stomp out. You didn't turn in your homework because your mom was sick and you had to help out extra at home. So when it comes to others, we blame the person. When it comes to our actions, we blame the situation.

What does this have to do with CFIT and controlling error? It has a lot to do with it. We just covered the four system-level failures that are often in place for a mishap to occur. We love to blame the situation for our problems. So it would be easy to say, "It's not my problem; it is a problem with the system." You could then proceed to blame organizational influences, unsafe supervision, or preconditions for unsafe acts—and you might be right. But remember, always remember, you the pilot (and the crew) are the last line of defense. If you hold, the mishap is likely to be avoided. That is our desire: to control error in order to prevent CFTT and CFIT. With that in mind, we turn our attention to what we can do to weaken or strengthen that last line of defense, you the pilot.

Attitudes

When it comes right down to it, error is often precipitated by how the pilot thinks. A pilot can reason effectively or ineffectively. Our reasoning, in turn, can be affected by our attitudes. There are numerous attitudes that we have outlined throughout this book that are unhealthy for the pilot. In the following, we'll quickly review those and add a few additional ones.

Hidden agendas

Our desire to accomplish a personal objective or hidden agenda can override sound judgment. Attitudes such as get-home-itis can cause us to manipulate the crew for our own ends. Another agenda is one-upmanship where we let another crew member or squadron mate persist in an error in order to make ourselves look better. We desire to look good through the contrast effect compared with the guy making the errors.

Arrogance through experience

High experience can make us proud and unable to take advice or listen to others. We become too good for our own good. Often the experienced pilot will rely on his or her heuristics, or rules of thumb, to get the pilot by. The heuristics almost become second nature. The pilot feels that they have proved reliable in the past, so there is no need to change now. But sometimes, we need to change. They say you can't teach an old dog new tricks, but it is to the dog's benefit to learn them.

Complacency

High experience can also cause us to become complacent. This complacency leads us to withhold the attention that aircraft flight deserves. Complacency can also be brought on by low workload or routine operations. Remember, complacency can lead to lost situational awareness.

Self-fulfilling prophecies

The expectations we have of others is translated in our speech and nonverbal body language. People often live up to or down to our expectations. If we expect high performance and enforce high standards, the rest of the crew often complies. However, if we treat people with disrespect, despise them, and have low expectations of their performance, they will often meet this expectation as well. This is known as a self-fulfilling prophecy; our actions and expectations actually bring an event to pass. Be careful of your expectations of your fellow crew members; they have an influence and are important!

Get-home-itis

This was already mentioned under hidden agendas, but it bears repeating. Pressing to get to your destination can be costly. Don't take unnecessary risks. Do a cost/benefit analysis. Are the benefits of trying to arrive at your destination worth the potential costs?

Antiauthority

Don't try to get back at the higher-ups by the way you fly. That is a poor idea. This attitude borders on that of the rogue pilot, the most dangerous pilot of all. For further reading on the rogue, see the fine book entitled *Darker Shades of Blue: The Rogue Pilot* by Tony Kern (1999).

Macho man

The macho pilot is the one who is trying to prove his manhood by the way he flies. This person usually has very low self-esteem and tries to compensate for that in some way. Macho pilots feel that they have to prove their prowess every time they fly. I had a full colonel wing commander in the Air Force who fit this mold. He was once flying the KC-135 (which he had little experience in) with an instructor pilot alongside. It was a day with heavy crosswinds at the runway. As they began their en route descent, the IP wisely and subtly asked, "Have you ever landed the Tanker in high crosswinds before?" "Oh yeah, sure, no problem" was the reply of the macho colonel. But the en route descent gave him plenty of time to think. Finally, the colonel asked, "Got any suggestions on this landing?" The IP now had the opportunity he had hoped for with the macho man.

Invulnerability

This malady strikes mainly the young: the young aviator who thinks he or she is Superman or Superwoman. Such pilots believe they are invincible and can do no wrong, thinking "it can only happen to the other guy." They are likely to press into danger, because they don't have enough sense to recognize it as danger. Remember, the superior pilot is one who uses superior judgment to avoid situations requiring the use of superior skills. We are all mortal. Sometimes the real old-head is also struck with symptoms of invulnerability. However, this is rare, as that pilot usually recognizes that he or she was fortunate to have escaped a few close calls. The close calls could come closer next time. Such pilots realize that they don't have unlimited "get out of jail free" cards.

Impulsiveness

This pilot acts before he or she thinks. I knew a pilot in a large cargo aircraft that once briefed crew members that if they got a fire light on takeoff, they would continue to "let it burn" until they were at least 2000 ft AGL. At 2000 ft they would then accomplish the appropriate boldface items and shut down the engine. They took off from a northern base very near max gross weight. Sure enough, they got a fire light on one of the four engines at about 400 ft AGL. The same pilot who had given the briefing reached up, and in a blink of an eye, he accomplished the boldface before the copilot could say "boo." The throttle was cut off and the fire shutoff handle pulled. He was supposed to confirm the engine before shutting it down as well, but he didn't. Fortunately, he pulled the correct engine. Because of his hasty action, the aircraft lumbered around the pattern just above rooftop level and was barely able to make it around to the touchdown zone of the runway. The rest of the crew was furious, to say the least. Impulsiveness is not an admirable trait in a pilot.

Resignation

This is the pilot who figures, what's the use? It doesn't matter, anyway. Nothing I do will make a difference. This, of course, is not true, as the United DC-10 crew that lost all flight control hydraulics aptly illustrates. That crew was able to handle a malfunction never before seen on the DC-10 and make an emergency landing at Sioux City with very little steering capability because it did not give up. You should always try to have an out. Keep thinking and never give up.

Air show syndrome

This type of pilot is the show-off. As Kern (1997) says, the two most dangerous words in aviation are "watch this." There have been many stories of pilots returning to their hometowns and putting on an impromptu air show for the locals. In an effort to wow their friends, they exceeded their personal and aircraft limits and ended up as lawn darts. This occurred with a T-37 crew in the South during the 1980s. If you want to show off, find another way to do it.

Emotional baggage

Also known as emotional jetlag, this is when we let a problem eat away at us. We can't let it go. We become so preoccupied with the mistake that we end up creating new mistakes through diverted attention. In economics this is known as "sunk costs." Economically we know that it is silly to keep throwing good money after bad. There comes a time when it is simply time to quit repairing the car with 250,000 mi on it and get a new one. As they say, you can't dwell on the past. You shouldn't cry over spilled milk. If you want to obsess about a problem, save it for when you land.

A related idea is when sometimes we feel as though something has been left undone. The answer is to do it so it stops gnawing at you or write it down with a grease pencil as a remainder to do it later.

Excessive professional deference

Respect for authority is good, but don't let it go too far. Your captain is not infallible. You are a valuable member of the crew. Your voice and input counts. Don't be afraid to speak up and ask questions. If your captain is an ogre, worry about the fallout later on. I realize that this is easier said than done, but as several accidents attest, it is vital to do. The Milgram obedience study showed us that people can be led to do things with simple commands. We must be mindful of that phenomenon. Two things can lead to excessive professional deference. The first is groupthink, or an overcommitment to group cohesion. Don't be so committed to group harmony that you are not willing to ask the important questions. The second is the halo effect. You may view your captain or flight lead with so much awe or respect because of his or her demonstrated abilities that you feel he or she can do no wrong. Or if the person does do wrong, you wonder who you are to point it out. Admiration of a flyer's expertise is laudable, but again, don't let it go too far. No one is perfect.

Copilot/passenger syndrome

This situation is where you are just along for the ride. When you are flying, you think, "The instructor will bail me out. She'll catch my mistakes." Fly as if you are the pilot in command. That will drive out the copilot syndrome.

Coke and peanuts syndrome

This is the flip side of the copilot syndrome. The Coke and peanuts syndrome is when you are the authority figure on the flight and/or the most experienced crew member. It is placing undue confidence in your fellow crew members—a confidence that is unbounded by reality. You don't know their actual abilities, but you assume they are up to speed. Ask the tough questions and ascertain the true state of your crew, especially crew members you have just met. Even if they are top-notch, don't place blind faith in their abilities. We are all prone to error.

Habit patterns

Habit patterns aren't truly attitudes, but they can reflect attitudes. Are your habit patterns marked by sloppiness? Or have you established good professional habit patterns? Remember, in a crisis situation, habit patterns can either save you or kill you.

Social loafing

This syndrome is related to the copilot syndrome listed previously. Social loafing is marked by not pulling your share of the load. You are content to let others do the work while you ride their coattails. This is an integral part of poor CRM. Social loafing leads to a failure to cross-check and back up. If you have two pilots and one of them is loafing, you essentially have a solo flight. The scary part is if the solo pilot doesn't realize he or she is solo and is relying on Mr. Loaf to be pulling his or her share. When it doesn't happen, disaster can follow.

Primary backup inversion

This occurs when we treat the backup system as the primary system—a role that system is not designed for. We may stop visually clearing for traffic because the TCAS will catch it. The primary system is your eyeball. The TCAS supplements your eyeballs; it doesn't replace them. Don't place the cart before the horse.

Time warps

Don't always assume you have all the time in the world. You don't. Earlier we mentioned that impulsiveness, on the other end of this spectrum, is not helpful either. Time can be compressed or elongated in our minds. What is important to realize is that it is good to set time boundaries for certain decisions and actions. Flight 173 into Portland did not set a time limit for dealing with the gear problem. As a result, it crash-landed because of fuel starvation. We can get caught up in the time trap while seeking the perfect solution. The perfect solution is one that maximizes safety, effectiveness, and efficiency and saves gas. However, we can't always get the perfect solution. Sometimes a workable solution is better than the best solution. Pursuing the best solution can sometimes make things worse. Of course, that is a hard pill to swallow for a perfectionist. A good solution in a timely manner is often what the pilot should strive for.

Ignorance

An attitude of arrogance can lead to an attitude of ignorance. Invulnerability can also lead to ignorance. It can cause us to ignore approach minimums. It can cause us to ignore approach procedures. It can cause us to ignore the need to conceive a backup plan. I think the two most annoying character traits are apathy and ignorance. They should not mark the pilot.

Errors in Thinking

I said earlier that the pilot (and crew) was the last line of defense in a system of error prevention. When it comes down to it, we are responsible for our flight and the errors we make. How we control error is influenced by our attitudes. Those attitudes affect our reasoning processes, our decision making, if you will. Our decision-making process is where the rubber meets the road, because clear thinking will usually lead to clear action. So to conclude this book, I want to focus on human reasoning ability. I think you will find as you read about human decision making that you see a correlation with many of the factors at work in the case studies covered in this book.

Norman's points

Donald Norman (1993) has done some excellent writing on human reasoning and how it relates to everyday life. These ideas relate directly to the cockpit. He notes that activities that rely upon a lot of human reasoning and planning run into problems because they rely on information inside the human being and that has some fundamental problems.

The first fundamental problem is that our information has a *lack of completeness*. You can never know everything that is relevant. Furthermore, our information has a *lack of precision.* We can't have accurate, precise information about every single relevant variable. Humans have an *inability to keep up with change.* The world is a very dynamic place. What holds at one moment may not hold in another. We may actually know the relevant information, but by the time we can act on it, the situation may have changed. If the problem at hand requires a *heavy memory load,* we run into problems. As mentioned previously, the capacity of our short-term memory is seven plus or minus two chunks of information. That is why we must use other crew members as stores of knowledge. We should view them as a library book that we can pull off the shelf for information. That is also why a grease pencil is the pilot's good friend. Writing information down eases our memory load. Even if we could remember everything, trying to recall it all would be a real bottleneck. Finally, some problems require a *heavy computational load.* If we could even know all the variables, accounting for them all would be a heavy burden. For example, the problem $2 \times 4 \times 8$ is a manageable computational load. But what if the problem were $248 \times 673 \times 318$? That is another matter indeed!

In the real world, because of all of the factors above, we experience time pressure, as we must often make decisions and take action quickly. This leads us to oversimplify our analysis of the problem, which in turn leads us to wrong actions. This is a gloomy picture. So what is the pilot to do? The best advice is to plan ahead and try to think through possible scenarios. Also, we should strive to be proficient at flying tasks so we can free up mental space to concentrate on the really tough problems. Finally, be flexible; we must respond to the situation and change activities as the world dictates. Remember, flexibility is the key to airpower.

Biases in thinking

Aside from Norman's helpful, but sobering, assessment of human reasoning, it is important to highlight several biases in thinking that lead to pilot error. As I review these, you should again recognize that these were illustrated in the case studies we have covered in this text. To help organize the information and orient the reader, I offer this simple model of how pilots make decisions. This model is influenced by the writings of Chris Wickens at the University of Illinois. The model of the pilot's decision-making process is illustrated as follows:

Stage 1: acquiring information (cue seeking)

Stage 2: assessing the situation (diagnosis)

Stage 3: choosing an action (decision)

As I list various biases and heuristics, be mindful of the distinction between the two. A *bias* is simply a way in which decision making is wrong. An example would be a conservative bias that every blip on the TCAS is a midair waiting to happen in crowded airspace. A *heuristic* is simply a mental shortcut or rule of thumb that helps us work more efficiently (but not always accurately). An example may be to assume a fuel flow of 10,000 lb/h during all cruise legs of the flight.

Stage 1 (cue seeking) factors

Salience bias   The salience bias is simply a pilot's tendency to give attention to whatever sticks out the most. An example would be the Eastern Airlines L10-11 that crashed into the Everglades. The crew devoted its attention to the two glowing green gear lights and the burned-out light because that is what caught its attention. It was so powerful that the crew ignored the C-chord that chimed when the altitude deviated by over 250 ft from the set altitude. The crew never hear that chord; it wasn't as salient as the gear lights.

Confirmation bias   This is the tendency to seek out information that confirms our hypothesis while ignoring evidence that is contradictory to our hypothesis. World War II ace Ben Drew relates a tragic but powerful illustration of the confirmation bias. He was flying his P-51 Mustang on a deep bomber escort mission during World War II as a member of a fighter formation. His flight lead was due to rotate back to the States and had yet to bag an aerial kill. It bothered the man because most of his squadron mates had at least one kill. Suddenly through the cloud deck came a formation. The flight lead ordered the aircraft to drop its tanks and pursue the bandits (enemy aircraft). A couple of the formation members weren't sure and questioned whether they might be just bogeys (unknown aircraft). The pilot barked, "They're bandits!" and jumped on one's tail and had him in a ball of flames before the pilot knew what had happened. The other bandit members turned for the fight. When they turned, it became clear that it was a formation of British Spitfires. The man had his air-to-air kill—of friendly forces. Drew points out that the combination of aggressiveness, expectations, and a stressful situational can be a setup for trouble. This Mustang pilot was under the confirmation bias, as is the pilot who flies into IMC because he or she is pretty sure of the existence of a hole that can be popped through. This pilot may choose to ignore the dark clouds on the other side of the hole—another example of the confirmation bias.

Channelized attention   Channelized attention (or cognitive tunneling or fixation) causes the pilot to seek cues in only one part of his or her environment. By focusing on only one cue, the pilot ignores other factors that may be important. This is closely related to the salience bias. Indeed, because a certain factor sticks out, the crew member may be drawn into channelized attention. Again, the classic example is the Eastern Airlines L10-11, whose crew fixated on a burned-out lightbulb as the plane slowly descended into the Everglades.

Perceptual sets Perceptual sets are mental models we slip into. We see the world in a certain way and it is hard to shake us out of it. Remember the crew that tried to land on 22L instead of 26L in Hawaii. It had discussed the fact that 22L was generally the landing runway and then ATIS confirmed that 22L was the landing runway. However, the approach controller cleared the crew to land on 26L, but to the aircraft commander's mind, the controller said 22L. Perceptual or mental sets can be very strong and heavily influence our perception of the world.

"As if" heuristic This is where a pilot treats ambiguous evidence as if it were conclusive and 100 percent reliable. We sometime don't like to deal with probabilistic reasoning (e.g., there is a 70 percent chance it will be VMC). We prefer to think of things in all-or-nothing (black-and-white) terms. That is why if the evidence is in favor of one position but not definitive, we will mentally treat it as if it is conclusive. This may lead us to treat an instrument that is "acting just a little bit funny" as if it is completely reliable.

Stage 2 (diagnosis) factors

Overconfidence bias The "as if" heuristic discussed above relates directly to the overconfidence bias. If we treat ambiguous information as if it were completely reliable, we may in turn place more confidence in the information than is warranted. We become too confident of our forecast or hypothesis. We are more confident of it than we should be. If we are overconfident, we may only give attention to factors that continue to support our position (the confirmation bias) and we may be less prepared for circumstances that will arise if our choices are wrong. As discussed earlier, this is much more likely in the highly experienced pilot.

Availability heuristic   When pilots assess a situation, they form hypotheses about what could be going on. Experienced pilots tend to form more hypotheses than newer pilots. Because they have more experience, they have more hypotheses to draw upon. Research shows that people are more likely to choose the hypotheses that can be most easily brought to mind. This is called the availability heuristic. The problem is, just because the hypothesis can be easily brought to mind doesn't mean that it is the most likely solution to the problem. The most available is not always the most likely. For example, if a pilot was recently briefed on a certain engine problem, the pilot may key on that problem when the engine starts acting up even if there is a far more likely explanation for the engine problem. The opposite also occurs. Things not easily brought to mind may be considered less likely to occur than they really are.

This problem can occur when deciding whether to risk a penetration into IMC as a VFR-only pilot. The pilot may think "Joe made it last week and I did it another time and it worked." What is most available to memory may not be the best course of action.

Strength of an idea   When we hold two competing beliefs simultaneously, cognitive dissonance occurs. *Cognitive dissonance* is simply that gnawing feeling that occurs when things don't quite add up. People in general don't like this tension. They seek to relieve the pressure by changing one of their beliefs or rationalizing their true action or belief away. Pilots are a microcosm of the person on the street. Cognitive dissonance is not valued on a flight deck. We want things to add up. We seek to rid the environment of that tension.

If that tension exists widespread in a cockpit, the crew may jump at the first idea that soothes the tension. This is especially true when it is a time-critical situation. It also helps if someone in authority makes the suggestion. The person benefits from the "halo effect" in that he or she is especially qualified to speak on flight matters. The problem is that the suggestion may not be a good one. However, crew members are afraid to question the idea because they feel much better since it was introduced. To challenge the idea would be to invite tension back into the aircraft. Therefore, the crew may limit itself in the number of alternatives available because it has a warm fuzzy from the first good sounding idea introduced.

Anchoring and adjustment This is the car salesperson routine. The salesperson gets you to focus on the retail price on the window sticker, then adjusts from that higher price to make you feel better. When it comes to pilots, once we anchor on something, we may be reluctant to give it up. For example, if we get a VMC weather forecast, we will be more likely to fly VFR into IMC because we figure it can't be that bad; after all, the weather was forecasted to be VMC.

Stage 3 (decision) factors

Simplifying bias As Don Norman pointed out, humans have a finite ability to deal with computational loads. Because of that we tend to simplify probabilistic information that comes out of Stage 2. We must combine this probabilistic information with the costs and the benefits of different outcomes. We may be trying to maximize things over the long run or trying to minimize poor outcomes in the short term. The captain of Flight 173 into Portland was struck with this bias. The captain had to weigh troubleshooting the gear and preparing the cabin against how much flight time the fuel reserves allowed. He chose to simply ignore the fuel, acting as if he had infinite flight time. If asked point-blank if he could fly indefinitely, he would say no. However, he treated the situation mentally as if he had indefinite reserves.

Availability heuristic (again) This heuristic also applies to the best course of action to take. A decision alternative that is available to memory because it has been chosen recently and has proven successful in the past is likely to be chosen again even if the current situational cues do not suggest it to be the wisest choice. This is one reason that the airlines and the military seek to standardize procedures for choices and decisions as much as possible. These procedures are often committed to memory so that they will be recalled first in the mind in the event of an emergency. This heuristic will also affect the estimation of risk on the part of the pilot. The pilot who has never had an accident may consider the probability of an accident as lower, thus underestimating his or her risk.

Framing bias This bias concerns the choice between an outcome that is uncertain (the risky alternative) and a certain outcome (the sure thing). If the situation is cast in negative terms, pilots tend to be risk seeking—for example, the pilot who must decide to cancel or delay a takeoff with a plane loaded with passengers due to "iffy weather." The pilot could also choose to go ahead and take off in the iffy conditions. There is a high probability of getting through the weather okay and a small chance of a crash or aircraft damage. However, if the pilot delays, that is a sure thing. The pilot will be late. When caught between a rock and a hard place (in the negative frame), the pilot will tend to be risk seeking and press ahead with the takeoff.

However, if you turn this situation around and frame it in a positive manner, the decision will likely change. Given the risky choice of taking off and getting to the destination on time (a good outcome) and the certain choice of safety (a good outcome), pilots will tend to be risk-aversive. They don't want to risk the positive sure thing. As you can see from this example, by simply framing the same situation in a certain light, the decision is altered 180 degrees. Crew members with a hidden agenda will often use this framing technique to their advantage.

Crew versus individual decisions

Judy Orasanu has done a lot of fine work in the area of crew decision making. Before leaving the topic of pilot decision making, it would be helpful to highlight some of the differences between crew decision making and individual decision making. She points out that crew decision making is managed decision making. The captain is responsible for making the decision but is supported by crew members (inside the cockpit), ATC, and ground crews (outside the cockpit).

Crews have the capability to make better decisions because they have multiple hands, ears, eyes, and so forth. This gives them increased mental horsepower. The crew can consider the bigger picture. Crew members have different perspectives to offer and a greater pool of information to draw on.

However, crews don't necessarily make better decisions. Their decisions can be worse. This can be caused by poor communication so that they don't understand the situation entirely or they don't understand the captain's intentions. An error can propagate through the crew, picking up validity as it goes. Crews then can grow too confident of how correct they are. Of course, crew members can also engage in social loafing, ducking their responsibilities, leaving the work for others, and causing interpersonal conflicts—or even fights.

The mentor

What this discussion should lead you to is the realization that good aircrew decision making is difficult. It is not impossible, but it is challenging. One thing that the pilot can do is to pick out a mentor. A mentor is simply a more experienced and respected pilot to learn from. Expert pilots have the advantage of having seen it all. They are familiar with the situations you may face, and they can help you identify these patterns. Once you learn to identify the patterns, you can learn the response or set of responses to match that pattern. An experienced pilot is a sage who can dispense knowledge if you are simply willing to pick his or her brain. This pilot can be invaluable in your development.

The experts

What does an expert decision maker or pilot look like? How can we identify one? What should we be shooting for? I think one answer lies in what we have learned about good decision makers in general. Research has revealed the following. Good decision makers tend to be better at task management and be more proactive, anticipating and preparing for future uncertainties. They are also better at time management. They tend to be more selective in applying different decision strategies given the nature of the problem. In other words, they are choosy. They shy away from the one simple approach used every time.

"When your only tool is a hammer, every problem is a nail"—that is not the motto of an expert. All of these characteristics can be translated directly to the pilot in the cockpit.

The Smokin' Hole Revisited

The Flight Safety Foundation has prepared an excellent CFIT prevention checklist (see the Appendix). It is a useful guide to help the pilot estimate his or her risk of CFIT. This is part of controlling error through preparation. Forewarned is forearmed. I encourage you to use this checklist as a preventative measure in your personal fight against CFTT and CFIT.

At the beginning of this book, I invited you to join me at the smokin' hole so that we could learn from war stories valuable lessons that can help us control error and avoid CFIT. My hope is that you've enjoyed your time here and that because of your time at *the* smokin' hole, you will not become *a* smokin' hole. The key is to learn and to *apply* what you have learned in flight. A great story is interesting to listen to, but if we don't learn from it, we have cheated ourselves. The storyteller often learns more than the listener. In this case, I know I have learned much, and I hope you can say the same. Live to learn and learn to live.

References and For Further Study

Hughes, R., R. Ginnett, and G. Curphy. 1999. *Leadership: Enhancing the Lessons of Experience,* 3d ed. Boston: Irwin/McGraw-Hill. Chap. 4.

Kern, T. 1999. *Darker Shades of Blue: The Rogue Pilot.* New York: McGraw-Hill.

Kern, T. 1997. *Redefining Airmanship.* New York: McGraw-Hill. Chap. 4.

Klien, G. A. 1993. A recognition-primed decision (RPD) model of rapid decision making. In *Decision Making in Action: Models and Methods.* Eds. G. Klein, J. Orasanu, R. Caulderwood, and C. Zsambok. Norwood, N.J.: Ablex.

Norman, D. A. 1993. *Things That Make Us Smart.* Reading, Mass.: Addison-Wesley. pp. 147–148.

O'Hare, D. and S. N. Roscoe. 1990. *Flightdeck Performance: The Human Factor.* Ames: Iowa State University Press.

Orasanu, J. 1993. Decision-making in the cockpit. In *Cockpit Resource Management.* Eds. E. L. Wiener, B. G. Kanki, and R. L. Helmreich. San Diego: Academic Press. pp. 137–172.

Orasanu, J. and U. Fischer. 1997. Finding decisions in natural environments: The view from the cockpit. In *Naturalistic Decision Making.* Eds. C. E. Zsambok and G. Klein. Mahwah, N.J.: Lawrence Erlbaum. pp. 343–357.

Reason, J. 1990. *Human Error.* Cambridge, England: Cambridge University Press.

Shappell, S. A. and D. A. Wiegmann. 2000. The human factors analysis and classification system—HFACS. DOT/FAA/AM-00/7.

Wickens, C. D. 1999. Cognitive factors in aviation. In *Handbook of Applied Cognition.* Eds. F. T. Durso, R. S. Nickerson, R. W. Schvaneveldt, S. T. Dumais, D. S. Lindsay, and M. T. H. Chi. New York: Wiley. pp. 247–282.

Wickens, C. D. and J. G. Hollands. 2000. *Engineering Psychology and Human Performance,* 3d ed. New York: Prentice-Hall.

Appendix:
CFIT Checklist*

*This checklist can be found on Flight Safety Foundation's web site (www.flightsafety.org/cfit.html).

Flight Safety Foundation

CFIT Checklist
Evaluate the Risk and Take Action

Flight Safety Foundation (FSF) designed this controlled-flight-into-terrain (CFIT) risk-assessment safety tool as part of its international program to reduce CFIT accidents, which present the greatest risks to aircraft, crews and passengers. The FSF CFIT Checklist is likely to undergo further developments, but the Foundation believes that the checklist is sufficiently developed to warrant distribution to the worldwide aviation community.

Use the checklist to evaluate specific flight operations and to enhance pilot awareness of the CFIT risk. The checklist is divided into three parts. In each part, numerical values are assigned to a variety of factors that the pilot/operator will use to score his/her own situation and to calculate a numerical total.

In *Part I: CFIT Risk Assessment*, the level of CFIT risk is calculated for each flight, sector or leg. In *Part II: CFIT Risk-reduction Factors*, Company Culture, Flight Standards, Hazard Awareness and Training, and Aircraft Equipment are factors, which are calculated in separate sections. In *Part III: Your CFIT Risk*, the totals of the four sections in *Part II* are combined into a single value (a positive number) and compared with the total (a negative number) in *Part I: CFIT Risk Assessment* to determine your CFIT Risk Score. To score the checklist, use a nonpermanent marker (do not use a ballpoint pen or pencil) and erase with a soft cloth.

Part I: CFIT Risk Assessment

Section 1 – Destination CFIT Risk Factors	Value	Score
Airport and Approach Control Capabilities:		
ATC approach radar with MSAWS	0	____
ATC minimum radar vectoring charts	0	____
ATC radar only	-10	____
ATC radar coverage limited by terrain masking	-15	____
No radar coverage available (out of service/not installed)	-30	____
No ATC service	-30	____
Expected Approach:		
Airport located in or near mountainous terrain	-20	____
ILS	0	____
VOR/DME	-15	____
Nonprecision approach with the approach slope from the FAF to the airport TD shallower than $2^3/_4$ degrees	-20	____
NDB	-30	____
Visual night "black-hole" approach	-30	____
Runway Lighting:		
Complete approach lighting system	0	____
Limited lighting system	-30	____
Controller/Pilot Language Skills:		
Controllers and pilots speak different primary languages	-20	____
Controllers' spoken English or ICAO phraseology poor	-20	____
Pilots' spoken English poor	-20	____
Departure:		
No published departure procedure	-10	____
Destination CFIT Risk Factors Total (−)		____

Section 2 – Risk Multiplier

	Value	Score
Your Company's Type of Operation (select only one value):		
Scheduled	1.0	
Nonscheduled	1.2	
Corporate	1.3	
Charter	1.5	
Business owner/pilot	2.0	
Regional	2.0	
Freight	2.5	
Domestic	1.0	
International	3.0	
Departure/Arrival Airport (select single highest applicable value):		
Australia/New Zealand	1.0	
United States/Canada	1.0	
Western Europe	1.3	
Middle East	1.1	
Southeast Asia	3.0	
Euro-Asia (Eastern Europe and Commonwealth of Independent States)	3.0	
South America/Caribbean	5.0	
Africa	8.0	
Weather/Night Conditions (select only one value):		
Night — no moon	2.0	
IMC	3.0	
Night and IMC	5.0	
Crew (select only one value):		
Single-pilot flight crew	1.5	
Flight crew duty day at maximum and ending with a night nonprecision approach	1.2	
Flight crew crosses five or more time zones	1.2	
Third day of multiple time-zone crossings	1.2	

Add Multiplier Values to Calculate Risk Multiplier Total ____

Destination CFIT Risk Factors Total × Risk Multiplier Total = CFIT Risk Factors Total (–) ____

Part II: CFIT Risk-reduction Factors

Section 1 – Company Culture

	Value	Score
Corporate/company management:		
Places safety before schedule	20	
CEO signs off on flight operations manual	20	
Maintains a centralized safety function	20	
Fosters reporting of all CFIT incidents without threat of discipline	20	
Fosters communication of hazards to others	15	
Requires standards for IFR currency and CRM training	15	
Places no negative connotation on a diversion or missed approach	20	

115-130 points	Tops in company culture	
105-115 points	Good, but not the best	**Company Culture Total (+)**____ *
80-105 points	Improvement needed	
Less than 80 points	High CFIT risk	

CFIT Checklist 341

Section 2 – Flight Standards

	Value	Score
Specific procedures are written for:		
Reviewing approach or departure procedures charts	10	___
Reviewing significant terrain along intended approach or departure course	20	___
Maximizing the use of ATC radar monitoring	10	___
Ensuring pilot(s) understand that ATC is using radar or radar coverage exists	20	___
Altitude changes	10	___
Ensuring checklist is complete before initiation of approach	10	___
Abbreviated checklist for missed approach	10	___
Briefing and observing MSA circles on approach charts as part of plate review	10	___
Checking crossing altitudes at IAF positions	10	___
Checking crossing altitudes at FAF and glideslope centering	10	___
Independent verification by PNF of minimum altitude during stepdown DME (VOR/DME or LOC/DME) approach	20	___
Requiring approach/departure procedure charts with terrain in color shaded contour formats	20	___
Radio-altitude setting and light-aural (below MDA) for backup on approach	10	___
Independent charts for both pilots, with adequate lighting and holders	10	___
Use of 500-foot altitude call and other enhanced procedures for NPA	10	___
Ensuring a sterile (free from distraction) cockpit, especially during IMC/night approach or departure	10	___
Crew rest, duty times and other considerations especially for multiple-time-zone operation	20	___
Periodic third-party or independent audit of procedures	10	___
Route and familiarization checks for new pilots		
Domestic	10	___
International	20	___
Airport familiarization aids, such as audiovisual aids	10	___
First officer to fly night or IMC approaches and the captain to monitor the approach	20	___
Jump-seat pilot (or engineer or mechanic) to help monitor terrain clearance and the approach in IMC or night conditions	20	___
Insisting that you fly the way that you train	25	___

300-335 points	Tops in CFIT flight standards	
270-300 points	Good, but not the best	**Flight Standards Total** (+) ___ •
200-270 points	Improvement needed	
Less than 200	High CFIT risk	

Section 3 – Hazard Awareness and Training

	Value	Score
Your company reviews training with the training department or training contractor	10	___
Your company's pilots are reviewed annually about the following:		
Flight standards operating procedures	20	___
Reasons for and examples of how the procedures can detect a CFIT "trap"	30	___
Recent and past CFIT incidents/accidents	50	___
Audiovisual aids to illustrate CFIT traps	50	___
Minimum altitude definitions for MORA, MOCA, MSA, MEA, etc.	15	___
You have a trained flight safety officer who rides the jump seat occasionally	25	___
You have flight safety periodicals that describe and analyze CFIT incidents	10	___
You have an incident/exceedance review and reporting program	20	___
Your organization investigates every instance in which minimum terrain clearance has been compromised	20	___

You annually practice recoveries from terrain with GPWS in the simulator........ 40 _____

You train the way that you fly... 25 _____

285-315 points	Tops in CFIT training	
250-285 points	Good, but not the best	**Hazard Awareness and Training Total** (+) _____ *
190-250 points	Improvement needed	
Less than 190	High CFIT risk	

Section 4 – Aircraft Equipment

	Value	Score

Aircraft includes:

Radio altimeter with cockpit display of full 2,500-foot range — captain only..... 20 _____

Radio altimeter with cockpit display of full 2,500-foot range — copilot............ 10 _____

First-generation GPWS... 20 _____

Second-generation GPWS or better.. 30 _____

GPWS with all approved modifications, data tables and service
bulletins to reduce false warnings... 10 _____

Navigation display and FMS.. 10 _____

Limited number of automated altitude callouts... 10 _____

Radio-altitude automated callouts for nonprecision
approach (not heard on ILS approach) and procedure.............................. 10 _____

Preselected radio altitudes to provide automated callouts that
would not be heard during normal nonprecision approach....................... 10 _____

Barometric altitudes and radio altitudes to give automated
"decision" or "minimums" callouts.. 10 _____

An automated excessive "bank angle" callout.. 10 _____

Auto flight/vertical speed mode.. -10 _____

Auto flight/vertical speed mode with no GPWS... -20 _____

GPS or other long-range navigation equipment to supplement
NDB-only approach... 15 _____

Terrain-navigation display.. 20 _____

Ground-mapping radar... 10 _____

175-195 points	Excellent equipment to minimize CFIT risk	
155-175 points	Good, but not the best	
115-155 points	Improvement needed	**Aircraft Equipment Total** (+) _____ *
Less than 115	High CFIT risk	

Company Culture _____ **+ Flight Standards** _____ **+ Hazard Awareness and Training** _____

+ Aircraft Equipment _____ **= CFIT Risk-reduction Factors Total** (+) _____

*** If any section in Part II scores less than "Good," a thorough review is warranted
of that aspect of the company's operation.**

Part III: Your CFIT Risk

Part I CFIT Risk Factors Total (–) _____ **+ Part II CFIT Risk-reduction Factors Total** (+) _____

= CFIT Risk Score (±) _____

**A negative CFIT Risk Score indicates a significant threat; review the sections in Part II and
determine what changes and improvements can be made to reduce CFIT risk.**

In the interest of aviation safety, this checklist may be reprinted in whole or in part, but credit must be given to
Flight Safety Foundation. To request more information or to offer comments about the FSF CFIT Checklist,
contact Robert H. Vandel, director of technical projects, Flight Safety Foundation, 601 Madison Street,
Suite 300, Alexandria, VA 22314 U.S., Phone: 703-739-6700 • Fax: 703-739-6708.

Index

About the Author

Daryl R. Smith, Ph.D., is a pilot, human factors engineer, aviation psychology professor, and lieutenant colonel at the United States Air Force Academy, where he teaches a number of performance-related courses to the next generation of warrior pilots. He has flown and instructed in multi-engine crewed aircraft as well as trainers. He resides near Colorado Springs with his wife Laura and children Ryan, Stephanie, and Andrew.

CPSIA information can be obtained at www.ICGtesting.com
Printed in the USA
LVOW132310190712

290838LV00008B/87/P